Westview Special Studies on China and East Asia

Food for One Billion: China's Agriculture Since 1949
Robert C. Hsu

This book examines the agricultural policies and programs adopted by the Chinese leadership since 1949 and analyzes the role of agriculture in China's changing development strategies. Dr. Hsu gives particular attention to the measures intended to improve agricultural technology and to the sources of funds for agricultural investment. He concludes that, although the collective system has been effective in mobilizing China's rural resources for agricultural development and in promoting progress in labor-intensive agricultural technology, periodic extreme leftist policies and interference by rural party cadres have caused various kinds of inefficiency, offsetting the advantages gained from collective farming.

This is the first book to systematically analyze the ways in which China's agricultural development is being financed. By critically examining the level and nature of state resources allocated to agriculture, the author challenges the view that China has pursued an agriculture-first strategy of economic development since the early 1960s.

Dr. Hsu is an associate professor of economics at Clark University. He was educated at the National Taiwan University and Atlanta University and earned a Ph.D. in economics at the University of California (Berkeley). In 1977–1978 he was a Visiting Research Fellow at the University of Lund (Sweden) and then took a research trip to China. In 1980, as a Fulbright scholar, he was a member of the U.S. delegation to the first U.S.-ASEAN (Association of Southeast Asian Nations) Dialogue Program. Dr. Hsu specializes in the field of economic development and has published articles in the *American Journal of Agricultural Economics, Asian Survey, Economic Development and Cultural Change, Journal of Political Economy,* the *New England Journal of Medicine,* and elsewhere.

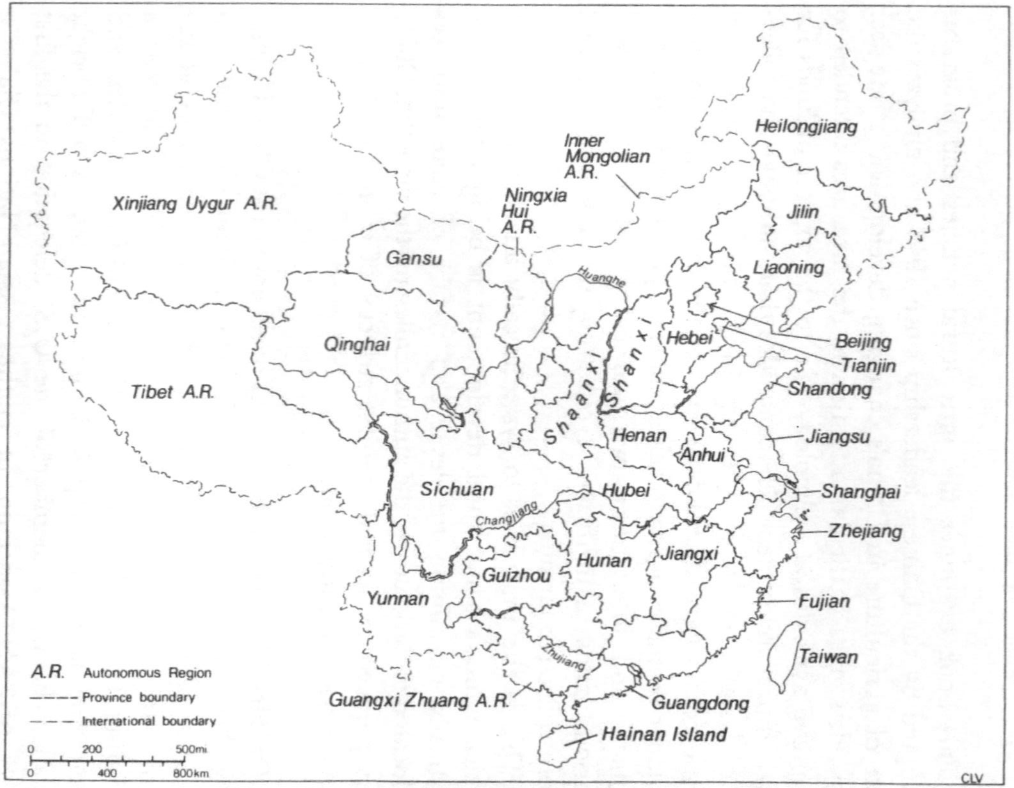

China. (Drawn by Clark University Cartography Laboratory)

Food for One Billion: China's Agriculture Since 1949

Robert C. Hsu

Routledge
Taylor & Francis Group
LONDON AND NEW YORK

First published 1982 by Westview Press, Inc.

Published 2018 by Routledge
52 Vanderbilt Avenue, New York, NY 10017
2 Park Square, Milton Park, Abingdon, Oxon OX14 4RN

Routledge is an imprint of the Taylor & Francis Group, an informa business

Library of Congress Cataloging in Publication Data
Hsu, Robert C.
Food for one billion.
(Westview special studies on China and East Asia)
Includes index.
1. Agriculture and state—China—History—20th century. 2. Agricultural innovations—China—History—20th century. I. Title. II. Series.
HD2098 1982 338.1'0951 82-6976
ISBN 13: 978-0-367-01840-5 (hbk) **AACR2**

To Sharon, Nancy, and Steven

Contents

Figures and Tables

Figures

Tables

Acknowledgments

This book is an outgrowth of my research for a course on the Chinese economy that I have been teaching at Clark University off and on since the early 1970s. The course and thus the book would not have been possible without the encouragement of my colleague, Roger Van Tassel, who saw the need for such a course as early as 1971.

The initial idea for the book came from Mervyn Adams Seldon, formerly of Westview Press, in 1979. Without her initiatives and enthusiasm, this book would not have been written. She has also given me helpful suggestions during the preparation of the manuscript.

In the course of my research and writing, I have also benefited from communications and discussions with other individuals. Nai-Ruenn Chen, whose insights into the Chinese economy have long been a source of inspiration to me, has made very valuable suggestions. Wen-yang Wen has greatly enhanced my understanding of the biological-chemical aspects of agricultural technology. Last, but not least, the intellectual curiosity of many students at Clark University has been a source of stimulation to me.

To all of these individuals go my heartfelt thanks. Naturally, all shortcomings of the book are my responsibility alone.

I would also like to take this opportunity to acknowledge the financial support that I have received over the years from various sources. This support has made my study of the Chinese economy possible and thus has contributed directly or indirectly to the writing of this book. A grant from the Clark University Faculty Research Fund made the initial research possible. A grant from the Svenska Handelsbanken for Social Science Research, Sweden, enabled me to study the science and technology policies of China at the Research Policy Program, University of Lund, Sweden, during the 1977–78 academic year. An international travel grant from the Mellon Foundation financed a study trip to China in the summer of 1978. All of these contributed greatly to my understanding of China's agriculture. Finally, the Clark University Research Fund financed the cartographic works for this book.

Robert C. Hsu

Introduction

The people of the United States tend to take food for granted. But to the majority of humanity, the daily struggle for food has always been the greatest challenge of life. In China, particularly, the people know only too well how bitter that struggle is, for no other country in the world has such an enormous task—feeding more than a billion people, almost one-quarter of the world's population, on 7 percent of the world's arable land.

This Malthusian nightmare of severe population pressure on limited land has haunted the Chinese for so long that it has affected their collective psyche, customs, and language. For example, a common form of greeting near mealtime is, "Have you eaten?" One's job is commonly referred to as one's "rice bowl." Interestingly, the coming of communism has even enriched this food-centered vocabulary and consciousness. For example, in recent years, the lifelong tenure system for party and government officials that has prevailed since 1949 has been criticized as providing an "iron rice bowl" to incompetent officials. Similarly, the egalitarian system of income distribution that prevailed in many places during the Cultural Revolution (1966–76) has been criticized for providing a "big rice pot" for everyone to eat from, and thus reducing people's incentives to work.

It is very important for the outside world to understand China's struggle to feed its population and to be supportive in their efforts. For one thing, a hungry people is an angry people, and the world cannot afford to see one-quarter of its population angry. Further, should China ever have to rely on outside food aid to feed its people, the availability of food aid to other poor countries will be greatly diminished. Similarly, if China has to import large volumes of food, world food prices will go up, causing difficulties for other import-dependent countries; China's ability to import Western technology and manufactured foods also will be adversely affected. Finally, China's experiences in various agricultural policies, rural organizations, and agricultural technology provide valuable lessons to other developing countries. China has made some costly mistakes, which other countries should learn to avoid. On the other hand, it has made remarkable progress in some areas, particularly in the area of labor-intensive agricultural technology, which should be of great value to other developing countries.

This book was writtten to introduce readers with little knowledge of China to a wide range of issues in China's agricultural development. The emphasis is on the period since the Communist Revolution in 1949, particularly the post-Mao period since 1977. However, to understand China's agriculture, one first of all has to appreciate the natural constraints imposed on it by China's geo-environmental conditions, such as topography, climate, and soils. These constraints cannot be violated without disastrous consequences. Thus they have influenced all aspects of China's agriculture, including cropping patterns, farming practices, and agricultural technology. Chapter 1 discusses these constraints as well as the consequences of violating them, as the Chinese have done in some cases.

Similarly, to understand China's agricultural policies and performance in the post-1949 period, one has to have some historical perspective. The Communists inherited the agricultural and other rural problems and the socioeconomic conditions that caused them. These problems, discussed in Chapter 2, constituted the initial conditions of the post-1949 period and conditioned to a large extent the range of policies that could be adopted to deal with them. In fact, vestiges of the past are still visible in rural China.

The bulk of the book is devoted to various policy issues, particularly issues in economic policy and agricultural technology. One of the major economic issues discussed in Chapter 3 concerns the peasants' incentives to produce. Since China's agriculture was collectivized in the 1950s, the Chinese leaders have oscillated in their efforts to mobilize the peasants between the leftist approach of emphasizing the socialist ideology, collective organization, and moral incentives (late 1950s, 1966–76) and the more liberal policies of giving the peasants more material incentives and economic freedom (early 1960s and since 1977). Evidence to date suggests that excessive reliance on the socialist ideology has not been effective and that the peasants are more interested in concrete material gains and welfare than in ideological abstractions.

This lesson has not been learned easily, for many lives have been needlessly lost and many resources wasted in the periodic ideological polemics and power struggles between proponents of conflicting policies, all in the name of building a socialist China. It is in light of this that policies of the current leadership are significant and augur well for China's future. As Deng Xiaoping, the prime mover of the post-Mao modernization program, once said, "Black cat or white cat, the one that catches mice is a good cat." Contrast this pragmatic emphasis on function and result with the radical leaders' obsessions with ideological purity during the Cultural Revolution lest the nation should "change color." As discussed in Chapters 3 and 7, during the Cultural Revolution much national energy was wasted in ideological excesses at all levels. For example, in the rural areas local party cadres were busily giving out orders prohibiting the peasants from growing fruit trees on their tiny private plots and from growing produce and making handicrafts for sale on the rural markets lest they should become rich and "restore capitalism."

The current leadership has concluded that valuable resources can be better spent on rewarding the "good cat," or productive peasants, and on promoting agricultural modernization than on such ideological purification. Thus, income distribution is to be in accordance with work performed. The peasants have more freedom in working on their private plots and in household sideline production, and rural free markets are reopened. Party cadres' political interference in agricultural production is to be minimized. Also, as shown in Chapter 6, state procurement prices for agricultural products have been raised, government investment in agriculture has been stepped up, and agricultural loans have been increased. All of these changes are aimed at raising the peasants' incomes and incentives to produce.

Issues in agricultural technology are examined in Chapters 4 and 5. In several areas of agricultural technology—fertilization, irrigation, plant breeding, agricultural mechanization, rural energy supply, and plant protection—the Chinese have, through self-reliant efforts, made important progress. In particular, China leads the developing world in the promotion and improvement of small hydropower stations, in biogas generation, and in integrated pest control. The technology advanced in these areas should be of great value to other developing countries. China's approach to promoting technological progress is worthy of close attention. By adopting a strategy of "walking on two legs"—that is, the simultaneous development of modern technology by state research institutions and modern industries and of labor-intensive or intermediate technology by the research network and rural industry of the people's communes—the pace of technological development has been accelerated. The farmers are provided with modern inputs that otherwise would not be available in many areas. The commune system is also conducive to the adoption of labor-intensive methods of production because the communes are responsible for the employment of their members. Partly for this reason, the pace of farm mechanization has not been very fast in spite of government promotion, because it creates unemployment for communities that have no alternative employment opportunities for their peasants.

As with economic policies, China's policies toward agricultural technology have also fluctuated over time. The current leadership has accorded technological progress in agriculture a top priority, with both basic and applied research being emphasized. Technical education in the rural areas is being stepped up and political indoctrination is being deemphasized. This is in sharp contrast with the "red-over-expert" policy of the Cultural Revolution period, which stressed political attitude and ideological correctness at the expense of technical expertise at all levels of the society. China is still suffering, however, from the consequences of "red-over-expert" policy, and the general level of scientific and technical knowledge in the rural areas remains very low.

The development and adoption of improved agricultural technology and inputs require much capital investment, as discussed in Chapter 6. In China, the communes themselves have financed the bulk of agricultural investment. In fact, the commune system of collective ownership and investment has facilitated the mobilization of agricultural surplus as well as rural labor for capital investment. On the other hand, although the Chinese leaders have recognized the importance of agriculture to the economy ("agriculture as the foundation of the economy") since the early 1960s, they have not devoted adequate state resources, in the forms of agricultural price increases, agricultural loans, state investment, and subsidies, to the agricultural sector. Other sectors of the economy have received the bulk of state resources. This imbalance is finally being redressed in the early 1980s.

A final issue concerns the overall performance of China's agriculture. In Chapter 7, the performance of China's agriculture is evaluated from various points of view, such as equity in food distribution, growth rates, and crop yields. It should be pointed out, however, that the agricultural issue cannot be separated from the larger issue of China's economic-political system itself because all major agricultural policies have been determined primarily by the party leadership and implemented under the supervision of party cadres. Thus the question naturally arises: Has this system of party control and central planning been conducive to China's agricultural development? To answer this question, various factors are taken into account in Chapter 7: party politics and the political environment, the competence of rural party cadres, and the orientation of the party leadership and the soundness of its policies.

It can be seen from this brief introduction that the issues in China's agriculture are complex and multifaceted, and the answers to them are by no means straightforward. China's agriculture has traversed an uneven and uncharted course of development, partly because of natural resource stringency and population pressure, and partly because of man-made problems. In the future, the resource stringency and population pressure will only intensify. On the other hand, the current leadership is committed to agricultural modernization as the top priority of its modernization program, and the policies adopted to date have been conducive to that end. It is impossible to predict the future outcome, however, because it will depend on many unpredictable factors, including the implementation of the policies as well as the political environment over an extended period of time.

In any case, China's experience in dealing with the various issues and problems in its agricultural development is both enlightening and fascinating to study. It reveals various facets of human nature under changing and trying conditions, the strengths and weaknesses of various policies and institutions, and the efficacy and consequences of alternative types of agricultural technology. In its enormity and complexity, China's never-ending struggle to feed its population and to wrest the maximum possible from the "good earth" is also unparalleled in history and elsewhere in the world. It behooves the rest of the world to be concerned about it and to seek to understand it.

1
The Natural Environment

China is the third largest country in the world, second only to the Soviet Union and Canada. It has a land mass of 9.6 million square kilometers, and is thus slightly larger than the United States or Western Europe. Located in East Asia, China measures over 5,500 kilometers from north to south (36 degrees of latitude) and over 5,000 kilometers from east to west (60 degrees of longitude).[1]

In spite of its large size, China's agricultural land is limited. More than two-thirds of the land area consists of mountains, and about 11 percent is desert. Half of the country is arid or semiarid. As a result, only 13 percent of the land area is arable. Currently no more than 11 percent of the land is cultivated, which constitutes about 7 percent of the world's total arable land.[2]

Topography, climate, and soil conditions are the major determinants of land utilization and agricultural cropping patterns in China, as shall be seen in this chapter. In some areas, however, human activities have modified to some extent what nature has provided.

Topography and Rivers

Topographically, China is divided into three regions, descending from the Qinghai-Tibet Plateau in the west to the eastern coastal area like a west-east staircase (see frontispiece, p. iv).

The Qinghai-Tibet Plateau, known as "the roof of the world," is the highest and largest plateau in the world. It averages 4,000 meters above sea level and covers an area of 2.2 million square kilometers, about 23 percent of China's total land area. It is sparsely populated and most of it is unsuited to agriculture.

To the north and east of the Qinghai-Tibet Plateau, the terrain drops precipitously to the second step of the staircase, which averages between 2,000 and 1,000 meters above sea level. It consists of the Junggar, the Tarim, and the Sichuan basins, and the Inner Mongolia, the Loess, and the Yunnan-Guizhou plateaus.

Further to the east, from the northern tip of Northeast China to Guangxi in the south, lies the third step of the staircase, where the altitude of the land is less than 500 meters above sea level. This is the heartland of China's agriculture and includes, from north to south, the Northeast China Plain, the North China Plain, the middle-lower Changjiang (Yangtze River) Plain, and the Zhujiang (Pearl River) Delta. These plains constitute 12 percent of the country's total land area. Scattered among them is hilly land. In addition to being China's main granary, this region also has the greatest concentration of people, with a population density ranging from 50 to more than 200 persons per square kilometer. Off the southeastern and southern coasts of this region are the two major islands of China: Taiwan to the southeast of Fujian Province and Hainan Island south of Guangdong Province and the Guangxi Zhuang Autonomous Region (AR).

Because of the west-east descending nature of China's topography, the major rivers generally flow eastward. The two major rivers—the Changjiang and the Huanghe (Yellow River)—dissect the eastern half of the country and affect agriculture profoundly but differently.

The Changjiang is the longest river in China and the third longest in the world, having a total length of 6,300 kilometers. It drains a basin of 1.8 million square kilometers, about 19 percent of China's total area. It is the major artery of water navigation in central and eastern China and provides abundant water, about 40 percent of the total surface water flow, to irrigate farmland in the middle-lower Changjiang Plain.

The Huanghe is the second longest river in China, with a total length of 5,464 kilometers and a catchment area of more than 750,000 square kilometers. It is also the most heavily silted river in the world (thus the name Yellow River) and has wrought havoc in its floodings over the centuries. As it passes through the Loess Plateau, heavy rains wash 1.6 billion metric tons of soil annnually into the river. About 400 million tons of soil are deposited along its lower reaches to raise its riverbed by about 10 centimeters each year. As a result, in some places the river is 3 to 5 meters above the surrounding land. In the 2,000 years before 1949, there were some 1,500 dike bursts and 26 violent changes in its course.[3]

Since the 1950s, many conservancy projects have been built to help control the river. However, because of increased soil erosion due to accelerated deforestation and clearing of grassland along its upper and middle reaches, the rate of silting in the river has increased from about 1.3 billion metric tons a year in the early 1950s to 1.6 billion tons in 1980.[4] Every cubic meter of its water carries 34 kilograms of silt.[5] This high silt content makes much of the water unsuitable for irrigation because the silt would clog the irrigation canals.

In the south, the Zhujiang is the longest river (2,100 kilometers). In the Zhujiang Delta it provides irrigation water, about 20 percent of China's total surface water flow, and facilitates water transport.

Climate

China's climate is characterized by great regional and seasonal variations, due to China's great range in latitude, its monsoonal winds, and its complex togography.

China's extensive territory encompasses the frigid, temperate, subtropical, and tropical zones. The southern part of Guangdong, Yunnan, and Taiwan have a tropical climate. Heilongjiang in the northeast has long and severe winters. In between, however, the bulk of China lies in the subtropical, warm-temperate, and temperate zones. The north-south temperature difference is very great in winter. For example, the January mean temperature ranges from −30°C in the extreme northeast to above 15°C in the southern coastal areas. The summer temperature difference is much smaller because the temperature for the whole country is relatively high. For example, the July mean temperature is about 28°C in the south and 20°C in the northeast.

Precipitation amounts also vary greatly from the south to the north. From more than 2,000 millimeters in the far south, annual precipitation diminishes to about 1,000 millimeters south of the Changjiang and on the Yunnan-Guizhou Plateau, and to between 600 and 800 millimeters in the Huanghe Valley. In the northeast, annual rainfall ranges from 400 to 1,000 millimeters.

Because China is situated at the eastern side of Asia and at the western edge of the Pacific, the interaction between the continental air mass and the oceanic air mass produces the monsoons. From October to April, high air pressure develops over Siberia as cold arctic air accumulates. As a result, the prevailing winter winds blow southward from Siberia, making a large part of China, especially North China, cold and dry. Only the southeastern coastal areas receive moisture from the tropical Pacific air mass.

From April to September, the reverse takes place. As the land mass is heated up, the air over the continent becomes warmer and thinner than that over the ocean. The moist, high-pressure Pacific air mass therefore moves toward the continent, bringing humidity and precipitation. About 86 percent of the annual precipitation of the country falls between April and September. The summer rainfall is heaviest over the southeastern provinces. As the monsoonal winds penetrate the land mass toward the northwest, the precipitation gradually decreases. In addition, the further inland one travels toward the northwest, the more the annual precipitation is concentrated in the summer season. These two factors make the north and northwest more susceptible to droughts and crop failures.

Altitude and terrain also cause climatic variations. Generally, at the same latitude, the temperature drops by 5°–6°C with every increase of 1,000 meters in altitude. Thus the temperature in the Tibet Plateau stays below 10°C even in July. The Sichuan Basin has a milder winter than regions of the same latitude in the middle and lower Changjiang Valley because it is protected by surrounding mountains from cold waves. The Qinling Mountains, the watershed between the Changjiang and the Huanghe valleys, keep cold northern air from moving further south during the winter and make it difficult for the ocean air to penetrate the northwest during the summer. Consequently, the Qinling Mountains constitute the natural dividing line between China's subtropical and warm-temperate zones. They also constitute an important dividing line for China's agricultural regions.

Soils

As in topography and climate, there are also great regional variations in soils in China. However, the different types of soils can be classified into two major groups. The first group comprises the moderately to strongly leached soils (the pedalfer soils) of the south, which are acid. The second group comprises the slightly leached and unleached calcium soils of the north (pedocals), which are rich in lime (calcium carbonate) and therefore are alkaline.[6] The two groups are separated roughly by a line in an east-west direction that passes through the Qinling Mountains. This line follows closely the northern boundary of the "rice region" in which rice cultivation predominates. South of the Qinling Mountains, the soils are acid. The further south one proceeds, the more acid the soil becomes due to increasing precipitation and leaching. Traveling from the southeast toward the northwest, one passes through nine major soil regions.

1. The tropical region of south Guangdong, south Guangxi, Hainan Island, and Taiwan have highly acid red and yellow lateritic soils. Because of heavy precipitation, most of the easily soluble nutrients are leached out, and constant application of fertilizer is essential for crop growth.

2. The area between the tropical region and the lower Changjiang Basin has red and yellow podzolic soils.

3. The lower Changjiang Basin is covered with yellow podzolic soils from deposits of alluvium.

4. Sichuan, east Shanxi, and Shandong have purple and brown forest soils. Leaching in this region is moderate and a balance between acidity and alkalinity is maintained.

5. The North China Plain has yellow, naturally fertile, alluvial soils from the Huanghe, Huaihe (Huai River) and Haihe (Hai River), and loess dust blown from the northwest.

6. West of the North China Plain is the Loess Plateau of Shanxi, Shaanxi, and Gansu, in which the loess soil is yellow and chestnut and very fertile.

7. In the northeast, most of the soils are podzols and brown forest soils. It is the major forest land of China.

8. From the western part of the northeast through eastern and

central Inner Mongolia are the steppes of China, with chernozem, chestnut, and brown soils. Agriculture is feasible, with irrigation, in these grasslands.

9. West of the loess land and the steppes are the great expanses of mountain desert, desert, and semidesert lands. Further northwest are the dryland and mountains of Northwest China. All these lands are dry, have some salt content, and are unsuitable for agriculture except in scattered oases where irrigation is possible. Approximately one-third of China's land is in these western and northwestern regions.

Land Utilization

Topography, climate, and soils are the major geo-environmental parameters that delineate the range of possibilities for land utilization and agricultural production. Within this range, the actual patterns of land use are determined by a number of factors, such as the food needs of the population, the level of agricultural technology, and industry's need for raw materials, which all change over time. The food needs of the population change as the size of the population and the standard of living of the society increase. The level of agricultural technology improves as a result of peasants' field experience and of research conducted in response to the agricultural needs of the society. Industry's demand for raw materials increases as industry expands. Consequently, land utilization and agricultural production will change over time, affecting the environment in the long run in many cases. For example, wasteland can be reclaimed for farming, rivers can be diverted for irritation, mountain slopes can be developed for grazing, and forests can be cleared for farming. Whether or not these changes are desirable for the society in the long run depends not only on the immediate economic cost-benefits but also on the long-term environmental consequences.

Of China's total land area, no more than 11 percent, or about 100 million hectares, are cultivated. About 80 percent of the cultivated land is planted to food crops. Of the land sown to food crops, rice paddy constitutes one-third and wheat one-fifth.[7] Other major food crops are maize (corn), gaoliang (sorghum), millet, and soybean. Among the industrial crops, the most important one is cotton, followed by tobacco, sugar, peanuts, and tea. China is the world's largest producer of tobacco.

The percentage of land cultivated is highest in North China where it exceeds 50 percent. A total of 47 million hectares of land, about 48 percent of the cultivated land, is irrigated, giving China the largest amount of irrigated land in the world. Most of the irrigated land is located in the south, reflecting the requirement of rice cultivation as well as the abundance of water in the south. The official target is to expand irrigated land to 60 million hectares by 1985.[8]

Between 1949 and 1978, about 20 million hectares of land were reclaimed for farming.[9] However, almost as much arable land has been lost to urban expansion, road and factory construction, and the encroachment of deserts. The prospects for sizable increases in the amount of cultivated land are not good. Although it is estimated that from 30 to 50 million hectares of wasteland, such as marshes, tideland, and hills, can still be reclaimed for farming,[10] problems of salinity and irrigation make such reclamation painfully slow and costly. Thus, the official target to reclaim over 13 million hectares of wasteland in the eight years between 1978 and 1985 seems overly ambitious.[11] Furthermore, the yield on the newly reclaimed land is invariably low, so that future increases in agricultural production have to come primarily from increases in yields, rather than increases in cultivated land, through the increased use of modern inputs, such as chemical fertilizer, improved seeds, and irrigation. These inputs will be discussed in Chapters 4 and 5.

About one-third of the land area consists of grassland suitable or potentially suitable for grazing. Most of it is located in the far northern and northwestern areas, the steppes, and in the eastern part of the Qinghai-Tibet Plateau. In general, China's grassland has not been fully developed or properly utilized. Some grassland has been turned into deserts or semideserts because of overgrazing without proper conservation, and some has been converted into farmland in the drive to achieve grain self-sufficiency in the pastoral areas. As a result, animal husbandry has been adversely affected.

Until the late 1970s, the Chinese government had not encouraged the development of commercial animal husbandry in the grassland to supply the heavily populated eastern areas. There are several reasons for this. First, except for a small percentage of the population—mostly the minority nationalities—the Chinese traditionally prefer pork to beef and mutton in their diet. Second, most of the grassland is located far from the eastern population centers, so the development of large-scale livestock production for the eastern markets involves transportation and refrigeration facilities, which have always been in short supply in China. On the other hand, hogs can be raised on small plots of land in the farming areas, thus minimizing the costs of transportation and storage. Third, there is a symbiotic relationship between farming and hog-raising—the peasants feed hogs food scraps and agricultural by-products and the hogs provide organic fertilizer for the farms. Consequently, pork accounts for more than 90 percent of the meat consumed in China, and it is produced primarily in the main agricultural regions.[12]

More than two-thirds of the land area consists of mountains, many of which are barren. Forests occupy only 12.7 percent of the total land area, a very low percentage compared with other countries.[13] Furthermore, it is very unevenly distributed, concentrated mostly in the northeast and in western Sichuan. In the northwestern provinces, forests constitute less than 1 percent of the land area.[14]

This paucity of forestland is the result of centuries of government neglect of proper forest management as well as indiscriminate deforestation by impoverished peasants to obtain firewood and lumber and to expand their farmland. Deforestation was accelerated in the late 1960s and throughout most of the 1970s to increase cropland for grain production. Between 1949 and 1978, 66.7 million hectares of land were afforested, but only one-third of this land escaped destruction.[15] The results of this deforestation are increased soil erosion and desert expansion, increased silting in the Huanghe and Changjiang,[16] greater temperature fluctuations and reduced rainfall in large areas, and a severe shortage of wood supply for the country. (These effects are discussed further in Chapters 4 and 5.) In 1979, new efforts were initiated to improve forest management and to increase afforestation, particularly in the north. The target is to increase forestland to 20 percent of the land area by the year 2000 and eventually to 30 percent.[17]

About 11 percent of the land area consists of deserts. Most of them lie in the north and northwest and a few in the northeast.[18] In general, China has had only limited success in controlling the spread of deserts. Since the early 1960s, an average of 1.3 million hectares of land a year in North China has been lost to deserts and there are still more than 40 million hectares of land that are vulnerable to desertization.[19] The spread of deserts is the result of inappropriate use of land in arid and semiarid areas, in flagrant disregard of the environment. Chief examples are the indiscriminate reclamation of wasteland to grow grains, excessive grazing, irrational use of water resources, and the felling of trees for firewood. Although measures have been taken since the late 1970s to halt this trend, effective large-scale actions are still lacking.

Agricultural Regions

China's main agricultural area lies in the eastern half of the country (see Fig. 1.1). It is further divided into two major agricultural regions: the southern rice region and the northern wheat region. The dividing line is the east-west line that passes through the Qinling Mountains and the Huaihe.

The rice region includes the middle and lower reaches of the Changjiang, the hilly areas and plains along the southeastern coast, the Sichuan Basin, the Yunnan-Guizhou Plateau, and Taiwan. It has a mostly subtropical climate, with a tropical climate in the far south. Annual precipitation is high, ranging from 1,000 millimeters to more than 2,000 millimeters. The soils are acid, and the frost-free growing season ranges from less than 270 days to as long as 365 days. The main crop is paddy rice grown on irrigated fields, and double cropping is extensive. In some places, three crops a year can be grown. The rice region is further divided into five areas.[20]

1. *Rice–winter wheat area*: most of Hubei and southern parts of Anhui and Jiangsu. Many important nonfood crops, such as cotton and mulberry trees, are also grown.

Figure 1.1. Agricultural regions of China. (Drawn by Clark University Cartography Laboratory)

2. *Rice-tea area*: most of Hunan, Jiangxi, Zhejiang, and Fujian.
3. *Sichuan rice area*: all of Sichuan and a small part of Hubei, Shanxi, and Gansu. Wheat, corn, rapeseed, and sugar cane are also grown.
4. *Double-crop rice area*: all of Guangdong and Taiwan, most of Guangxi, and southern parts of Fujian and Jiangxi. Three crops a year can be grown in some places. Sweet potatoes and sugar cane are also important crops.
5. *Southwestern rice area*: all of Yunnan and Guizhou, and the western part of Guangxi.

The wheat region includes the North China Plain, the Loess Plateau, the southern section of the Inner Mongolian Plateau, and the Gansu Corridor. The region has a temperate to warm-temperate climate. Precipitation is seasonal and inadequate, averaging about 500 millimeters a year. The soils are alkaline, and the growing season ranges from less than 180 days to about 240 days. Consequently, dry farming prevails, with comparatively little irrigation. Wheat and coarse grains (millet, corn, and gaoliang) are the major crops of the region. One crop a year or three crops every two years are grown north of the Great Wall, and three crops every two years or two crops a year are grown south of the Great Wall. Within the region, three areas can be distinguished according to the dominant cropping pattern.

1. *Spring wheat area*: northern parts of Hebei, Shaanxi, Shanxi, and Gansu, and the southern parts of Ningxia and Inner Mongolia.
2. *Millet-corn-winter wheat area*: most of Shanxi, Shaanxi, and Gansu, and small parts of Henan and Hebei. This is the Loess Plateau. Cotton is also grown.
3. *Corn-gaoliang-winter wheat area*: all of Shandong, a large part of Hebei and Henan, and northern parts of Jiangsu and Anhui, that is, the whole of the North China Plain. Rice, millet, cotton, and soybean are also grown.

Outside the main agricultural regions, there are also four agricultural areas.

1. *Corn–gaoliang–soybean–spring wheat area*: the three northeastern provinces of Liaoning, Jilin, and Heilongjiang, that is, the whole of the Northeast China Plain. The growing season is from less than 100 days to about 180 days.
2. *Mongolian pastoral area*: most of Inner Mongolia.
3. *The pasture land of Tibet.*
4. *Oasis farming area*: Xinjiang.

The Northeast China Plain is becoming a major grain-producing area because of its relatively high level of farm mechanization and increasing crop yields. Many large mechanized state farms are located in the area, particularly in Heilongjiang. Inner Mongolia, Tibet, and Xinjiang are agriculturally insignificant.

2
Historical Background

Early Development of Agriculture in China

China's agriculture dates back to the early and middle Neolithic period, that is, at least 8,000 or 9,000 years ago. However, the nature of its origin is still a subject of debate and reappraisal. On the basis of archaeological finds made in the North China Plain in the 1920s and 1930s, some scholars have long believed that Chinese agriculture and civilization originated in the flood plain of the lower Huanghe. The river provided abundant water and fertile soil, making the area suited to irrigated agriculture.[1] Consequently, irrigation was considered crucial to the rise of China's early agriculture and civilization.[2]

New archaeological and other scientific findings made since the 1950s suggest an alternative theory—that the southeastern part of the semiarid Loess Plateau in North China is the cradle of China's agriculture and civilization, and that ancient Chinese agriculture was predominantly dryland farming during the first four millenia of its history.[3] According to this theory, the Loess Plateau is suited to grain-centered agriculture because, in spite of its semiarid climate, the loess in China is alkaline, porous, and pliable, and thus fertile and easy to cultivate with primitive wooden digging sticks. The ancient Chinese farmers adapted their agriculture to the limited rainfall by growing drought-resistant varieties of cereals such as millet and sorghum. The river meant, to them, a threat of flooding, and thus something to be feared. It was not until the third century B.C. that large-scale irrigation networks appeared in China.[4]

In any case, millet and sorghum were indigenous to China and were adapted to the semiarid climate of the Loess Plateau. Wheat and barley were introduced into North China from Southwest

Asia during the latter half of the second millenium B.C.[5] As they required more irrigation, they were not popular as dryland crops but became important crops in the lower Huanghe Plain. Rice was traditionally believed to have been introduced from India into China. However, there is evidence that it was cultivated in North China around 3000 B.C. or earlier and that the early varieties could be indigenous to China.[6] By the Han dynasty (206 B.C.–221 A.D.), rice was cultivated as a wetfield crop in both North China and the Changjiang Valley.[7] The technique of transplanting seedlings from the nursery to the field, a method still used in China, was developed to accommodate the shorter growing season in North China and maximize the utilization of land.[8]

By the end of the Zhou dynasty (1134–256 B.C.), the entire North China Plain was fully settled. With the expansion of the nation and the growth of the population, new agricultural regions were gradually developed and settled throughout China's history. The new regions were the Changjiang Valley, settled by the Han dynasty; the southern areas, during the Tang (618–907) and Song (947–1279) dynasties; and the southwest and northeast, by the Qing dynasty (1644–1911). As a result, total cultivated land increased. Estimates of the cultivated area from 1400 to 1933 are given in Table 2.1. It can be seen that the expansion in total cultivated area began to slow down in the mid- to late nineteenth century after the last arable frontier in the northeast was settled.

With the migration of the population and the settlement of new land, superior agricultural technology was diffused from the developed to the undeveloped regions. In addition, the new environment often stimulated creative responses to the challenges it presented, leading to the invention of new technology, which stimulated agricultural development. Ever since its early periods, China's agriculture had experienced important inventions and technological improvements. For example, archaeological finds indicate that plows of bone or wood were used as early as 5000 B.C.[9] Written records indicate that iron implements, draft animals, fertilizers, and irrigation were in use during the Warring States period (403–221 B.C.). Improved iron plows, drawn by either oxen or men, were widely used during the Han dynasty[10] (Fig. 2.1). Throughout the subsequent dynasties, irrigation and water control were important areas of public works and were par-

TABLE 2.1
Population and Cultivated Area Estimates, 1400-1933

Year	Population (millions)	Cultivated Area[a] (million hectares)
1400	65-80	25 (±5)
1600	120-200	33 (±7)
1770	270 (±25)	63 (±7)
1850	410 (±25)	n.a.
1873	350 (±25)	81 (±3)
1893	385 (±25)	83 (±3)
1913	430 (±25)	91 (±3)
1933	500 (±25)	98 (±3)

Source: Dwight H. Perkins, Agricultural Development in China, 1368-1968 (Chicago: Aldine Publishing Co., 1969), p. 16.

n.a. = not available.

[a]Includes all land on which crops are grown, but excludes pasture-land.

ticularly essential for the expansion of rice cultivation in the south.

Better farming techniques were developed and refined by the peasants through centuries of field experience. From as early as 200 B.C., scholars and officials who had observed or had personal experience in farming often recorded the prevailing farming techniques. By the sixth century A.D. there had accumulated such a wealth of technical information on various aspects of agriculture as to make possible the compilation of the first comprehensive agricultural encyclopedia.[11] As new inventions were made and better farming practices developed in subsequent centuries, other books on agriculture were published to help disseminate these innovations. The invention of woodblock printing in the ninth century further facilitated this dissemination process.

Between the eighth and twelfth centuries, profound technological development revolutionized China's agriculture and by the thirteenth century gave China what one economic historian described as "probably the most sophisticated agriculture in the world."[12] First came the development of better milling machinery, which led to the widespread switching from the cultivation of millet to that of wheat in the north. Then came the development

Figure 2.1. Farmer in ancient times using an ox-drawn iron plow. Rubbing from ancient tomb, ca. Han Dynasty (206 B.C.–A.D. 221). The design has not changed much over the centuries. (From the author's collection)

and diffusion in the south of techniques of wetfield rice cultivation such as the dam, the sluice-gate, the noria (peripheral pot-wheel), and the treadle water-pump. Associated with these major inventions were better techniques to prepare the soil by using improved plows and by more extensive application of organic fertilizer.[13]

Finally, there was the introduction of better seeds and new crops, which improved land utilization and increased food production. In the eleventh century, during the Song dynasty, early-ripening Champa rice was introduced from Indochina and was widely adopted. It made double cropping possible in China's southern rice areas. In addition, Champa rice was relatively drought resistant and could be grown on poor land where other rice varieties could not.[14] Other new rice strains were created by selective breeding and were so widely adopted that by the eleventh century most of the strains in use in the eighth century had disappeared.[15] In the sixteenth century, New World crops such as maize, sweet potato, and peanuts were introduced, and later the Irish potato. These new crops enabled the Chinese to cultivate dry hills, mountains, and sandy loams to support a growing population.[16]

These technological innovations and improvements led to higher crop yields. In particular, the innovations between the eighth and twelfth centuries were the essential elements of what has been described by economic historians as the "medieval economic revolution" in China.[17]

After the thirteenth century, the pace of agricultural innovation slowed down. Nevertheless, China's agricultural yields continued to increase due to the diffusion of existing superior farming practices and techniques from the advanced regions to the lagging ones. Thus, as early as the seventeenth century, rice yield reached 2.3 metric tons per hectare,[18] a figure that many South and Southeast Asian countries had not reached in 1979 (see Table 7.2). As a result of the increase in yield and cultivated area, total agricultural output rose steadily. It has been estimated that between the late fourteenth and early nineteenth centuries China's grain output increased five to six times. About half of the increase came from expanded cultivated land and the other half from increases in yield per unit of land. This increased grain output enabled the population to grow at about the same rate during that period.[19]

The Late Qing Dynasty (19th Century to 1911)

By the nineteenth century, the expansion of cultivated land had slowed. Also, there were no technological innovations made in the nineteenth century. Furthermore, because of the dynastic decline during the period in general and the Civil War in the mid-century in particular, the number of water-control projects drastically declined.

As shown in Table 2.1, in the late nineteenth century, the amount of cultivated land barely increased; it was about 81 million hectares in 1873 and 83 million hectares in 1893, an increase of only 2.5 percent in two decades. Population declined from 410 million in 1850 to 350 million in 1873 due to the Civil War, then increased to 385 million by 1893. According to estimates, the size of the average farm in China declined from 1.37 hectares in 1870 to 1.35 hectares in 1890 and 1.06 hectares in 1910. The size of the average farm in the wheat region was 1.77 hectares in 1890 and 1.32 hectares in 1910. In the rice region, it was 0.81 hectare in 1890 and 0.77 hectare in 1910.[20]

In agricultural technology, no indigenous innovations were made. However, efforts were made by the Qing government in the late nineteenth century to introduce Western agricultural technology as part of the "self-strengthening" efforts to revive the declining empire and to resist the growing Western encroachment on China. First, a number of modern agricultural schools were established. The initial one was the Sericultural School in Hangzhou, Jejiang, in 1897. The Imperial Beijing University, established in 1898, also included an agricultural department. Experts from Japan and the United States, as well as Chinese experts trained in these two countries, served as teachers in these schools. Books and journals from these two countries were also imported. In some areas, private agricultural associations were formed for the purpose of learning and propagating new agricultural knowledge.[21] Second, a number of agricultural experimental stations, operated by the agricultural schools, were set up in the provinces between 1903 and 1911. The Central Agricultural Experiment Station in Beijing was established in 1906.[22] Finally, in the late nineteenth century and the early 1900s, some farmers began to apply chemical fertilizers. Three chemical fertilizer plants were established between 1905 and 1908 in Shandong.[23]

With the exception of the Sericultural School, which succeeded in introducing superior Japanese methods to several provinces, these changes did not produce noticeable effects on agricultural production because of the limited scale and the lack of competent personnel. Chemical fertilizers were not widely used because the prices were too high. Thus, agricultural technology in the farming areas remained unchanged.

In spite of these beginnings in agricultural changes, the policy of the Qing government toward agriculture remained exploitative in nature, as it had been throughout the dynasties. Revenues from the land tax amounted to about 30 percent of the revenues of the central government in the early 1890s, but the central government spent less than 2 percent of its expenditures on public works that directly benefited agriculture.[24]

To offset the decline in the size of the farm and the stagnant agricultural technology, some peasants shifted from barley, gaoliang, and millet to higher-yielding and more labor-intensive food crops, such as corn, sweet potatoes, and sesame, and to high-income cash crops, such as cotton, tea, silk, and sugar cane. Superior varieties of cotton and tobacco were introduced from the United States.[25] These changes enabled total agricultural production to keep pace with population growth in spite of unfavorable conditions.

The Republican Period (1912 to 1948)

Cultivated land expanded from 91 million hectares in 1913 to 98 million hectares in 1933, an increase of 8 percent in two decades. About 70 percent of the increase was in the northeast where soybean and gaoliang, among other crops, were planted on the new land.[26] Population increased at a faster rate, from 430 million in 1913 to 500 million in 1933, or an increase of 14 percent. As a result, the land-population ratio continued to deteriorate. It is estimated that the size of the average farm in the wheat region decreased from 1.32 hectares in 1910 to 1.10 hectares in 1930; in the rice region it declined from 0.77 to 0.72 hectare in the same period.[27]

In terms of technological changes in agriculture, the beginnings made by the Qing government in research and education were expanded. Thus, by the late 1940s, there were over 100 agricultural experiment stations at the national and provincial levels, 25 national agricultural colleges, and 9 agricultural schools. Cooperation with foreign institutions, such as Cornell University and the Rockefeller Foundation, was initiated in the 1920 and 1930s, resulting in the introduction of modern plant breeding science and the use of insecticides in 1935–40.[28]

In general, however, these educational-experimental efforts were too small in scale to affect productivity. Experiment stations at the provincial level were generally of poor quality due to lack of competent personnel and lack of support by local governments. The greatest deficiency, however, was the lack of an agricultural extension service; government efforts did not go beyond the planning stage.[29]

The single most important factor hampering agricultural development during the Republican period was the protracted political instability and the resultant squandering of resources on military expenditures. The period from 1912 to 1928 saw incessant rivalries among regional warlords backed by the Western powers and Japan. Between 1930 and 1935, military campaigns by the Nationalist government against the communist insurgents intensified. Between 1928 and 1937, with the exception of 1930, annual military expenditures consistently exceeded 40 percent of the national government expenditures. The second largest expenditure item, loans and indemnity service, took 25–38 percent of the budget.[30] Thus, very little government money was left for economic development. Furthermore, in allocating the meager resources left for the economy, agriculture was consistently given a low priority by the government.

To counteract the effects of rising population pressure on the land, stagnant technology, and government neglect, the peasants had no choice but to cultivate their land more intensively with more labor. However, the increased application of labor within the traditional technology had reached the point of diminishing returns. Average rice yield reached only 2.47 metric tons per hectare in the 1930s, which, although not low by international standards, was only 7 percent higher than that of the seventeenth century.[31] The average grain output per capita was very low, about

1.4 metric tons, as compared with 20 metric tons in the United States in the 1930s.[32]

Fortunately, the market conditions for agricultural products before 1931 were favorable, due to rising demand for exports and for urban industries. Some peasants took advantage of this by shifting to high-income cash crops. Between 1931 and 1935, however, the Chinese farmers experienced falling prices for their products because of (1) declining export demand brought about by the world depression and (2) the outflow of silver from China due to higher gold price of silver.[33] This further reduced the already low level of farm income.

The low agricultural output per capita left the economy with little margin of surplus above the minimum level of consumption. A natural disaster or a war could easily be devastating to the population. The poorer segment of the rural population in particular was in a very precarious situation because of the unequal distribution of land ownership and farm income. For example, from 1929 to 1933, Buck estimated that 33 percent of the farmers were landless tenants and 23 percent did not have enough land and had to rent.[34] Another survey showed that in the early 1930s, the landlords, who constituted 3 percent of the rural families, owned 26 percent of the land; on the other hand, the bottom 68 percent of the families owned 22 percent of the land.[35] In the early 1930s, from 39 percent to 56 percent of the rural families incurred debt to meet household consumption needs.[36]

To help alleviate the hardship of the rural tenants, a land law was passed in 1930 by the Nationalist government to reduce the rents. However, the law was never seriously implemented by the government because of political opposition. One prominent government official explained the lack of government action by saying that

> [During the 1930s] we . . . did not attempt to put a special check on great landlords. A more truthful way of putting it is that fundamentally we did not pay any attention to this problem. This was because we believed that in the Chinese ethical society the landlord and his tenants lived together like members of the same family.[37]

Many scholars would dispute the last point. For example, Tawney observed in 1932 that "a large proportion of Chinese peasants are constantly on the brink of actual destitution . . . the position of the rural population [in some districts] is that of a man standing permanently up to the neck in water, so that even a ripple is sufficient to drown him."[38]

The Sino-Japanese War of 1937–45 greatly disrupted China's agriculture and the rest of the economy. After the war, the Civil War between the Nationalists and the Communists erupted, which culminated in the Communist victory in 1949. The increasing hardships and poverty of the rural peasantry and the longstanding neglect of rural problems by the Nationalist government contributed in no small measure to that Communist victory.

Causes of Technological Stagnation and Population Growth

In this chapter we have seen that China's agriculture had an early period of technological innovation through the twelfth century, but afterward gradually became technologically stagnant. As Tawney aptly put it, China's peasants "ploughed with iron when Europe used wood, and continued to plough with it when Europe used steel."[39] The question is: Why?

There is no consensus among scholars on the answer,[40] and it is beyond the scope of this book to investigate the question thoroughly. However, some elements of the answer can be outlined. An important element lies in the interaction between food production and population growth. Increased food production in traditional China was associated with population growth rather than with increased surplus for investment and industrialization, as was the case in Western Europe in the seventeenth and eighteenth centuries. Population growth reduced the margin of surplus for investment and innovations. Labor was abundant so, rationally, farmers used more labor and economized on the use of expensive nonlabor inputs. At the same time, population growth necessitated continuously increased food production. This could be accomplished by expanding land used for cultivation or by raising the yield per unit of land. As arable land in the old settled regions became exhausted, migration to new regions took place. Grain yield per unit of land could be increased by the peasants working longer and harder—that is, by applying more of their labor to various tasks, such as better soil preparation, selecting better seeds, more weeding, improving on simple tools, using more organic fertilizers to improve the soil, and making use of available water. Through trial and error, efficient methods of accomplishing these tasks were found and disseminated.

These technological improvements were incremental in nature. As China has a long history of agriculture, many of the innovations were made early in its history. However, the point was gradually reached where further advances of a similar nature became increasingly difficult. In other words, the limits of labor-intensive, field-experience-based technology were gradually approached. Further advances had to take the form of more basic changes or breakthroughs with the help of modern physical, biological, chemical, and mechanical sciences. However, the development of these natural sciences requires the presence of two minimum conditions: an intellectual environment that allows people to be inquisitive about the physical world and a sizable economic surplus that can be devoted to inquiries about the physical world. By and large, traditional China lacked both of these requirements.

In the intellectual environment, the educational-intellectual-philosophical orientation of the traditional Chinese society since Confucius had always been humanistic rather than scientific—concerned with art, literature, ethics, and human relations rather than with systematic knowledge of the physical world. Technical expertise and experimentation were never accorded the respect given to the scholar-gentleman. During the Han dynasty, Confucianism was adopted as the state ideology. This intellectual orthodoxy, reinforced politically by the unified empire, became more monolithic and was not conducive to the diversity of thoughts and inquiries that was essential to the rise of sciences. This was especially true after 1300 when the neo-Confucian literati turned to introspection and intuition, or the mind and the spirit, even to explain matter or physical phenomena.[41]

In any country an economic surplus is needed for investment and for the support of technological-scientific pursuits. The size of the economic surplus in China was relatively low for the size of the country, compared with premodern Europe, the main reason being that income was distributed relatively equally in traditional

China compared with premodern Europe. Most of the potential economic surplus went to support population growth rather than investment. There were three socioeconomic reasons for this. First, aside from the imperial clans, traditional China "did not have a European-style hereditary superstructure (of which medieval church was a part) to produce lasting, if not permanent, pockets of wealth."[42] Second, due to the equal inheritance system that prevailed in China since primogeniture was abolished more than 2000 years ago, family wealth was equally divided among the sons and thus was more rapidly dissipated over time. The Chinese saying that "family wealth does not go beyond three generations" had much validity in traditional China. In rural areas, the equal inheritance system further fragmented land holdings but permitted more families to be engaged in farming. At the same time, it reduced the surplus attainable from the land. It also intensified the pressure to increase crop yield per unit of land by applying more labor and by making labor-using and land- and capital-saving innovations. Third, although land ownership in traditional China was far from equal, relatively favorable tenancy and rent conditions permitted tenants to share a large part of farm output.[43]

Given the orientation of the intellectual environment in traditional China, the low level of investable surplus due to a relatively equal distribution of income led to two consequences. The government and the wealthy families would not be inclined to support scientific-technological pursuits that were outside the mainstream of China's culture. In agriculture, farmers attempting to increase crop yield had to rely on their own field experience, personal ingenuity, and meager resources to make small technical improvements. Also, the relatively equal distribution of income made it possible to spread increases in food production thinly to support a larger number of people. As a result, the population continued to grow, and once the early and relatively easy phase of labor-intensive innovations was over, technological stagnation in agriculture began to set in.

3
Agriculture in China's Development Strategies

When the Chinese Communists came to power in 1949, they inherited a war-torn and underdeveloped economy with a population of about 540 million. Agriculture was characterized by low productivity, having virtually exhausted all possibilities of increase within the traditional technology. To ease the population pressure on the land and to develop the economy, the government had to give top priority to the development of agriculture. At an early stage of economic development, agriculture is critical to the rest of the economy, particularly fledgling industry, in the following ways.[1]

- It produces food for the population and raw materials for industry.
- It can provide economic surplus in the forms of savings and taxes for investment in the rest of the economy.
- It often produces the only exportable products to earn the foreign exchanges needed to import machines and technology for industrial production.
- It stimulates industrial production by providing the rural markets for manufactured goods.
- It provides employment to the bulk of the population. As agricultural productivity increases and industry expands, some of the labor engaged in agricultural production can then be released for industrial employment.

China's new leaders were aware of the important role of agriculture in the economy. However, there was no consensus among them as to the best way to develop it. Furthermore, in actual policies and in the allocation of state resources, they have not always accorded agriculture the priority it deserves. As a result, China's agriculture since 1949 has gone through a tortuous course of development, causing various problems for the economy, as discussed in this chapter.

Land Reform (1950 to 1952)

The first step taken by the new leadership in 1950 to develop agriculture was to carry out land reform. However, this was not merely an economic measure to redistribute land in order to relieve the impoverished peasants and to stimulate agricultural production. It was part of the larger political and social revolution to destroy the traditional rural order and the rural elite so that a new socialist order could be constructed and the Communist political power consolidated. The ultimate goal of the government was to socialize agriculture and to raise agricultural output.[2]

In the Chinese Communist lexicon, China's traditional rural class structure consisted of landlords, rich peasants, middle peasants, poor peasants, and laborers. The landlords did little or no work and drew their incomes from rents and usury. The rich peasants worked on their land but had to hire labor because they had more land than they could cultivate. The middle peasants neither hired labor nor worked for others. The poor peasants did not have enough land of their own. The laborers worked for others. The Communist leaders' tactics in the rural areas were to rely on the poor peasants, unite with the middle peasants, neutralize the rich peasants, and eliminate the landlords. Thus, in the land reform, the landlords' houses and land were confiscated and distributed, so much per capita, among the landless peasants and the landlords themselves. The rich and middle peasants were allowed to retain whatever they had. Many landlords who had abused their tenants were publicly tried and often subjected to violence.[3] The traditional landed gentry was destroyed, economically, socially, and politically.

The land reform affected agricultural efficiency and production in various ways. In terms of economic incentives, the ownership of land by the previously landless peasants must have had a favorable incentive effect on their production. Also, land reform drastically altered the rural distribution of income. The peasants did not have to pay for the land they received; nor were the landlords compensated for the land taken away from them. The rural economic surplus that was previously collected by the landlords in the form of rents was now partly retained by the peasants and partly collected by the state in the form of agricultural taxes. Finally, in terms of the scale and efficiency of farm production, the land reform created many more small farms and aggravated the fragmentation of land.[4] Many ex-tenants also lacked farm tools, which were previously provided by the landlords.

As the population pressure on the land was more serious than ever with the cessation of war, it was clear that more than land reform was needed to increase agricultural output. This additional step was to take the form of agricultural collectivization.

The First Five-Year Plan and Agricultural Collectivization (1953 to 1957)

After the land reform was completed in 1952, the Chinese leadership embarked on various ambitious development programs to attain the goals of rapid economic development and the building of socialism. During the First Five-Year Plan (1953–57), China sought assistance from the Soviet Union for the following reasons: (1) politically and ideologically, the Western countries were hostile to the new Chinese government; (2) there was a trade embargo imposed on China by the United States and some other U.N. members because of China's assistance to North Korea in the Korean War, and (3) the Soviet Union was the first socialist country to become successfully industrialized and was sympathetic to China's new government.

Consequently, China turned to the Soviet leaders for economic assistance and to the Soviet experience as a model in designing its development strategy. During this period, Soviet-designed projects formed the core of China's industrial program and constituted 48 percent of the planned industrial investment. In addition, between 1950 and 1960, some 11,000 Soviet specialists provided technical assistance in China. At the same time, some 28,000 Chinese were sent to the Soviet Union for training.

As a result, China's First Five-Year Plan was based on the

Soviet model of development. It can be characterized as a strategy of unbalanced growth, with top investment priority given to industry, especially heavy industry, and low priority given to agriculture. Industry received 52.4 percent of total investment, of which 89 percent was for heavy industry and only 7.8 percent was for agriculture.[5]

Although agriculture received few investment funds during the First Five-Year Plan period, great organizational efforts were made by the Chinese Communist Party, at Chairman Mao Zedong's insistence, to collectivize agriculture. Mao favored rapid collectivization, as opposed to a gradual approach advocated by some other leaders, because he was convinced that collectivization was a precondition for agricultural modernization in China. He believed it would be easier for the collectives to control the peasants' consumption and extract more economic surplus to finance agricultural modernization, and that modern agriculture works best on large collective farms. He advocated that China's agriculture be collectivized as soon as possible—before the peasants became accustomed to private ownership of land and aspired to become rich.[6]

Thus the First Five-Year Plan period saw the rapid transformation of individual farms into collective farms. The transformation went through three stages, each increasingly more collective.

1. At first, mutual aid teams were organized. These were teams of 6 to 15 neighboring households that pooled labor, draft animals, and farm tools together. The ownership of animals, tools, and land remained in the hands of individual households.

2. Elementary agricultural producers' cooperatives were organized in 1953–54. They consisted of 20 to 40 households. The ownership of land, farm tools, and animals was still private but their use and management lay with a management committee. This form of centralized management made it possible to undertake modest projects, such as irrigation ditches, pumping stations, and some mechanized farming, which would have been difficult for individual peasants to undertake. Net income, after taxes and expenses were deducted, was distributed to the members according to the amount of land and labor that they contributed.

3. Advanced agricultural producers' cooperatives were orga-

nized in 1956. They were formed by merging several elementary cooperatives and consisted of 100 to 300 households, the equivalent of a large village. Both the ownership and use of land and other capital goods were collective. The distribution of income was determined on the basis of labor contribution alone. In this way, the category of rich peasants was eliminated.

By 1957, the collectivization of China's agriculture was almost complete. Thus, within a span of five years, China's rural population was organized from individual peasants into members of some 752,000 advanced cooperatives, and the peasants' farms were transformed into collective farms. One important exception was that, during most of this period, private plots were allocated to peasant households. The peasants were allowed to devote their spare time to growing subsidiary crops or raising animals on these plots. The area of these private plots varied with time and region, ranging from 2 to 5 percent of the arable land per capita. However, in 1956, during the high tide of collectivization, the size of the private plot was usually less than the prescribed limit. Overzealous cadres in some areas disregarded the regulation and either completely abolished the plot or discouraged the peasants from working on it.[7]

All these institutional changes in agriculture brought about by collectivization had profound political and social consequences, such as the blurring of class distinctions. It is also of interest to note that China's collectivization was achieved relatively peacefully, in sharp contrast to the violent Soviet collectivization in 1929.[8]

The strategy of unbalanced growth in favor of industry during the First Five-Year Plan was based on the assumption that agriculture, once socialized, would generate enough investable surplus not only for its own needs but also for supporting industrial investment. Unfortunately, the anticipated surplus did not materialize in the 1950s for three reasons. First, as discussed in Chapters 1 and 2, China's arable land is extremely limited relative to its large population. Without large investment and modern technology, the low land-population ratio could not produce much beyond what was needed for minimum food consumption. Second, the collectivization of agriculture initially reduced the peasants' incentives to produce. Some peasants killed pigs and neglected the proper care of farm tools and animals in anticipation of collectivization. Finally, collectivization gave increased administrative power in the rural collectives to party cadres who knew little about farming and often made technical errors.

Not surprisingly, toward the end of the First Five-Year Plan, various problems of imbalance in the economy emerged. The low level of investment in agriculture led to slower growth in agriculture than in industry. This created shortages in some raw materials, such as cotton for industry, causing inflationary pressure and excess industrial capacity. Also, with the population growing at an annual rate of 2.3 percent, the slow growth in grain production meant that China's balance between food supply and demand was very precarious and there was no reserve against crop failure. The slow growth of agriculture also meant that many of the new plants had to operate below capacity for lack of rural demand.

There was also the problem of rural unemployment or underemployment. The Chinese estimated that, in 1958, from 10 to 20 percent of the rural labor force in some localities was surplus.[9] The capital-intensive nature of the Soviet model of development generated little industrial employment opportunities for the population. The Second Five-Year Plan, which was eventually shelved but which would have continued the Soviet model, envisaged an increase in nonagricultural employment of only seven million persons between 1958 and 1962, which was roughly equal to the annual increment in the labor force.[10] On the other hand, if labor-intensive plants were built in rural areas, rural unemployment problems could be solved or alleviated without much capital investment. The plants would also help raise the productivity of agriculture. The resultant increase in income would stimulate urban industry by increasing the demand for its products.

These considerations prompted Mao Zedong to adopt a new strategy of development in 1958 and to launch the Great Leap Forward to implement it.

The Great Leap Forward (1958 to 1960) and the People's Communes

The objective of the Great Leap Forward was to mobilize rural resources to accelerate the development of the traditional sector—that is, agriculture and light industry—so that balanced growth between the modern and the traditional sectors could be restored. It was expected that with such balanced growth, maximum economic growth would be attained. An annual rate of industrial growth of 25 percent or more for three years was envisaged and thus the term *Great Leap Forward.* This strategy of the simultaneous development of the modern and traditional sectors was called "walking on two legs." The rationale for it was that "if there were only one leg, that is heavy industry, without the other leg, that is, agriculture and light industry, or if the other leg is too short, it will be impossible to develop the national economy at top speed."[11] The programs to develop agriculture and light industry included the small industry campaign and various rural irrigation projects, which helped to launch the establishment of the people's communes.

The small industry campaign aimed at establishing small, labor-intensive plants in the countryside to help agriculture and light industry. It utilized indigenous, labor-intensive methods of production so that employment of rural manpower could be maximized. The capital and raw materials needed for the plants were to come from the local communities, thus bringing about some degree of decentralization not seen in the First Five-Year Plan. These plants included iron and steel foundries, fertilizer factories, oil extraction plants, machine shops, cement manufacturing facilities, coal and iron ore mines, and food processing plants. Most of them catered to the needs of the local farming communities. In the first nine months of 1958, 7.5 million plants were established, most of which were small handicraft workshops operated by the people's communes.

In addition to these rural industries, the government also encouraged the construction of a large number of irrigation projects by the agricultural collectives in 1957–58. The objective was to expand irrigated land by mobilizing idle rural labor, especially during the winter. Because the larger irrigation projects covered areas that included several advanced agricultural producers' cooperatives, special waterwork brigades were organized that drew manpower from the several cooperatives. These brigades were mobile units, moved from one construction site to another, and were organized according to rational division of labor. This method of organization was then increasingly applied to farm work. However, it became apparent that the advanced cooperatives were too small to allow for large specialized work brigades and for operating the small rural industries. At the same time, a labor shortage was developing because of the water brigades and small industries. To cope with these problems, some areas started amalgamating several advanced agricultural producers' cooperatives into a single commune in April 1958.

These communes met with the party leaders' blessings because, by making it easier to mobilize rural labor, the communes seemed to provide the institutional basis for higher agricultural productivity. In August 1958, the top leadership of the Chinese Communist Party adopted a resolution to transform the advanced cooperatives into communes. By the end of 1958, it was estimated that 99 percent of the peasant households were members of 26,400 communes.

The communes in 1958 differed from the advanced cooperatives in many respects. The communes were much larger, consisting of 4,000 to 5,000 households as compared with 100 to 300 in an advanced cooperative. Also, a commune was an administrative unit, corresponding to a township, within which farming, industrial production, commerce, and construction were carried out. The commune authority had the power to allocate labor and other resources within its vast area. The distribution of income in a commune was based partly on work and partly on free supply of food, and was therefore more egalitarian than under the cooperatives. The free supply of food was subsequently discontinued because it reduced the peasants' work incentives. At first, the peasants' private plots were totally abolished, as were rural free markets.

Although the communes have gone through frequent changes since 1958, including changes in their number (from 26,400 in 1958 to 50,000 in the late 1970s), some organizational features of the early communes have remained intact. The most important of these is the organization into three levels within a commune—production team, production brigade, and commune—each having well-defined functions.

The production team, consisting of from 15 to 50 households, is the basic level of collective agriculture. Most of the farm tasks are carried out by the production team. Team members are assigned work tasks and awarded "work points" for the work completed. These work points are the basis for income distribution.

The production brigade, with from 15 to 50 constituent teams, supervises the teams and coordinates their activities. It also organizes small-scale rural enterprises and capital projects.

The commune organizes larger rural enterprises and capital projects, and provides commune-wide services. At the same time, being the lowest level of government in the countryside, it is responsible for the implementation of government policies in its area. A commune may consist of from several to more than a dozen production brigades.

The rationale of the Great Leap Forward—to use labor-intensive rural projects to help develop agriculture—was sound in theory since China had a shortage of capital and an abundance of labor. In reality, however, these projects contributed little to agricultural development in the late 1950s and might have contributed to the agricultural crisis of 1959–61 for the following reasons.

First, there were problems of inefficiency. Most small plants were established hastily under political pressure during the campaign, without proper planning and cost analysis. Often the communities did not have skilled workers to operate these plants. Furthermore, their small scale caused inefficient operation. As a result, the products were often of inferior quality and much labor and materials were wasted. The "backyard furnace" movement to make iron and steel by the indigenous method was the most notorious example.

Second, there were problems of mismanagement and decentralization. As the rural industries were primarily the responsibility of the local communities and were established in a period of great political zeal with "politics in command," control of rural industries fell into the hands of party cadres. Most of the cadres lacked technical training and experience and tended to downgrade technical expertise. Many ill-conceived plants were set up that not only were inefficient but also competed with state projects for materials, manpower, and funds.

Finally, there were problems of political interference and mismanagement in farming and in the construction of irrigation projects. The establishment of the people's communes in 1958 gave much decision-making power to party cadres in the communes in

TABLE 3.1
Grain and Cotton Production, 1949-80
(million metric tons)

Year	Grain Production	Cotton Production	Year	Grain Production	Cotton Production
1949	111	0.4	1965	194	1.9
1950	130	0.7	1966	215	1.8
1951	141	1.0	1967	225	1.9
1952	161	1.3	1968	210	1.8
1953	164	1.2	1969	215	1.8
1954	166	1.1	1970	243	2.0
1955	180	1.5	1971	246	2.2
1956	188	1.5	1972	240	2.1
1957	191	1.6	1973	266	2.5
1958	206	1.7	1974	275	2.5
1959	171	1.2	1975	284	2.4
1960	159	0.9	1976	285	2.3
1961	168	0.8	1977	283	2.0
1962	180	1.0	1978	305	2.2
1963	190	1.2	1979	332	2.2
1964	194	1.7	1980	318	2.7

Sources: Joint Economic Committee, Chinese Economy Post-Mao (Washington, D.C.: U.S. Government Printing Office, 1978), p. 649 for 1949-76; Beijing Review, No. 27, July 6, 1979, p. 38 for 1977-78; ibid., No. 19, May 12, 1980, p. 14 & No. 19, May 11, 1981, p. 25 for 1979-80.

spite of their lack of farming experience. In their zeal to increase agricultural output, which was an index of their performance, party cadres often made technological blunders. For example, excessive deep plowing was ordered by cadres in some areas. Many poorly designed irrigation projects without proper drainage were constructed at the insistence of cadres, making large areas vulnerable to flooding and salinization. The peasants were "mobilized" on a large scale to carry out the cadres' orders to the point of physical exhaustion and antipathy; mass mobilization degenerated into "commandism."[12]

In 1959, there was abnormally heavy rainfall, damaging many of the newly constructed dams, canals, and reservoirs, and reducing cultivated land. The abnormal weather lasted for three years, causing great crop damage. Grain output sharply declined from 206 million metric tons in 1958 to 171 million in 1959 and 159 million in 1960 (see Table 3.1). Foodstuff had to be rationed, and large quantities of grain were imported from abroad in 1961 to relieve the food shortage.

The agricultural crisis had an adverse impact on industrial output because the supply of raw materials was reduced and state investment in industry was cut back. To compound the problems, the Sino-Soviet dispute reached a new height. Soviet technicians working in China on aid projects were abruptly withdrawn in 1960, taking their blueprints with them and leaving many projects unfinished. Thus, in both agriculture and industry the impetus and programs of the Great Leap Forward quickly came to an end in 1960. What followed were salvage operations to rescue the economy from total collapse.

Recovery and Adjustment (1961 to 1965)

After the failure of the Great Leap Forward, the immediate goal of the government policy was survival—to salvage the economy from the fiasco of the Great Leap and from natural disasters. Various policy changes were introduced. To alleviate food shortages, China started importing large quantities of wheat in 1961. The wheat-import program has continued since then, with annual imports averaging about 4–5 million metric tons.[13] Ideology was soft pedaled in favor of pragmatic measures, with more emphasis on economic incentives and technical expertise. The order of priority in the economy was completely reversed. Agriculture was now the "foundation" of the economy, that is, the top priority. After agriculture came light industry, and then heavy industry.

The new strategy of "taking agriculture as the foundation and industry as the leading factor" was originally propounded by Mao Zedong and was accepted by other members of the party leadership. Views differed, however, on how to implement the strategy. Two major opposing views can be distinguished. Liu Shaoqi, then Chairman of the Republic and vice-chairman of the Chinese Communist Party, argued that the new strategy required not only a large investment in agriculture but also more emphasis on the material incentives of the peasants and less stress on the socialist, collectivist organizations and activities in the rural areas. Mao agreed on the need for large investment in agriculture but stressed the need for more socialist education. He was opposed to capitalistic measures of material incentives for fear of the restoration of capitalism.[14]

Mao's influence suffered from the reverses of the Great Leap, whereas Liu had the support of many government officials; therefore, actual economic policies adopted during 1961–65 were made by Liu and his supporters in the government. Changes were made in agricultural policies aimed at stimulating the peasants' incentives to produce and at upgrading agricultural technology. The incentives included the restoration of private plots and rural free markets. The size of the communes was reduced to increase management efficiency. The basic ownership and accounting level in the communes was transferred downward from the commune to the production team so that the peasants' work and reward could be more closely related. Individual peasant households were even assigned a state production quota and given independent responsibility or a contract for fulfilling the quota. Surplus over the quota belonged to the peasants.

Significant progress was made in various areas of agricultural technology as more attention was paid to planning, quality control, and economic calculations. The quality of water conservation projects was improved over a wide area. As a result, by January 1966, one-quarter or about 6.25 million hectares of China's paddy fields had been transformed into high-yield fields with effective irrigation and drainage.[15] Mechanical innovation centered around semimechanization and tool improvement appropriate to China's conditions. Rural electrification was also stepped up. State investment in both irrigation projects and semimechanization concentrated on the high-yield farms of the south where water supply was abundant and, consequently, investment return was higher. A new priority was given to the use of agricultural chemicals, especially chemical fertilizers, to raise yields. The production of chemical fertilizers became the priority of industry, and increased attention was paid to the development of new strains of rice that were more responsive to these chemical fertilizers.

The Cultural Revolution Decade (1966 to 1976)

There was a Third Five-Year Plan for 1966–70, which would have developed the Chinese economy along the lines of the first half of the 1960s, but at an accelerated rate. The prospects for the plan were auspicious in 1966, as the economy had successfully recovered from the Great Leap and was on its way to rapid growth. However, by the mid-1960s, some of the policies responsible for the successful recovery of the economy were causing concern to Mao and his followers. In their view, the nation under Liu Shaoqi was increasingly deviating from socialism toward capitalism, and dedication to the revolution had given way to pursuit of material gains—peasants were devoting more and more time to their private plots, rural markets were flourishing, and workers and managers in factories had to be motivated by bonuses. In short, China was well on the road to the "revisionism" of the Soviet Union and Yugoslavia, which was anathema to Mao.

To reverse this trend, Mao launched the Cultural Revolution in 1966 to overthrow Liu Shaoqui. As Liu had the support of a well-established party and government bureaucracy, Mao had to rely on Lin Biao, the defense minister, an ad hoc Cultural Revolution Group headed by Jiang Qing, Mao's wife, and millions of idealistic Red Guards in his attempt to oust Liu from power. The ensuing power struggle and political instability lasted for an entire decade until Mao died and the radical Gang of Four was arrested in 1976. The decade of the Cultural Revolution saw turmoils that rivaled those of the Great Leap Forward. Both movements were highly ideological, launched by Mao and guided by his vision of a classless socialist China. The Third Five-Year Plan was vitiated by the Cultural Revolution as the Second Five-Year Plan was by the Great Leap.

Mao's success in the purging of Liu and his supporters from the party and the government in the late 1960s led to the dominance of extreme leftist ideology and policies in China through 1976, affecting profoundly all sectors of the society. The long-term development of science and technology in China was greatly retarded by the radical ideology and policies in two ways.

First, during the late 1960s, schools and universities were closed for from one to four years. Teachers, if not persecuted and jailed, were engaged, like their students, in political activities. The resultant loss of scientists, doctors, technicians, and teachers has been estimated at over a million. When schools and universities were reopened in the late 1960s and early 1970s, students were admitted on the basis of their political attitudes rather than on the basis of their academic qualifications. Thus the quality of the students was inferior. The quality of education was also lowered: the length of study was shortened and the curriculum was changed to emphasize political studies and practical courses. Basic sciences, essential for research in agricultural technology, were neglected.

Second, the policies toward science and technology were profoundly altered. Scientific studies not couched in Marxist-Maoist terms were regarded as bourgeois or reactionary. Research not immediately geared to production problems was regarded as elitist and irrelevant. Many scientists and experts from bourgeois backgrounds were persecuted and sent to labor farms or jails. The new philosophy of science was that "all innovations are created by the laborers in their production struggle and scientific experiments" and that "the humblest are the most intelligent." In accordance with this proletarian philosophy, "mass science" was promoted and glorified to encourage peasants and workers to solve production problems, and professional science was downgraded. Therefore, although small incremental improvements were made by the peasants in their farming activities, China lagged further behind other countries in the basic research essential for major advances in agricultural technology.[16]

The same extreme leftist ideology and policies also resulted in increased political interference with peasants' incentives and farming activities. Private household economic activities were viewed as conducive to the restoration of capitalism. Many communes were at times ordered by local party leaders to abrogate private plots or to set severe restrictions on the use of private plots. The scope of permitted household sideline activities was also severely restricted. For example, in some communes the peasants were prohibited from growing vegetables and fruit trees on the private plots.[17] In one extreme case, the raising of sows by peasant households was criticized as "supplying oxygen and blood to capitalism,"[18] although it was permissible to raise hogs. The reasoning presumably was that sows would multiply rapidly and make the family sideline a capitalist enterprise.

Egalitarianism was stressed to the extent of hurting the peasants' incentives. The radical ideology was that the distribution of income in accordance with labor was "work-pointism" or "work-point in command," an obsession with "material stimulation," which was to be discouraged. The Dazai brigade, hailed as a model in socialist farming throughout the decade, practiced a system of income distribution based on self-evaluation of work performance subject to group approval. As it had an egalitarian tendency for lack of objective performance criteria, it was recommended as a model to emulate. The Dazai brigade also had the production brigade rather than the production team as the level of accounting. This higher level of collectivity was considered to be more in line with socialism.

Although Mao had stressed the development of a diversified agriculture ("Take grain as the key link, and promote all-rounded development in agriculture"), his overzealous supporters distorted it to require agricultural regions to be self-sufficient in grain production irrespective of their natural conditions. Thus some forestland in the northwest and northeast and some grassland in the northwest were indiscriminately converted into farmland at the command of local party leaders in order to increase grain production, with the result that the environment was adversely affected. Some collectives that had planted nonfood crops for reasons of season or local conditions were ordered to pull up the seedlings and plant food crops, although the seasons or conditions were not appropriate for the latter.[19] Regions that had long planted sugarcane or cotton had to alter their time-honored cropping pattern in order to comply with the party line.

This does not mean that there was no progress made in agricultural development during the Cultural Revolution decade. In fact, throughout the decade the official policy continued to regard agriculture as the "foundation" of the economy. Small but steady progress was made in agricultural technology, particularly in 1970–73 when Zhou Enlai, the pragmatic premier, was able to moderate the excesses of the extreme leftists. Research in applied agricultural technology progressed. The use of better inputs, such as improved seeds, chemical fertilizer, insecticide, and better farm implements, steadily increased. The use of medium and small farm machines, such as hand-guided tractors and pumps, was promoted.

Agricultural technology and development also benefited from the importation of Western technology in spite of the stress on self-reliance and the overall slow growth in foreign trade during the decade. In particular, 13 large urea plants were purchased from Japan and West Germany during 1972–74, which greatly expanded the capacity of the chemical fertilizer industry.

Finally, an integral part of the strategy to develop agriculture was the revival of the rural industries. Thousands of small rural plants were built by the county governments and the rural collectives. These plants were agriculture oriented; that is, they either produced products for agricultural use or processed agricultural products. The objectives were to raise agricultural productivity and to increase rural employment. However, problems of inefficiency and high cost due to the small scale of production and poor management plagued many rural plants. These problems were aggravated by the same political-ideological interference that the rural collectives encountered in farming. Party cadres often restricted the scope and the nature of the industrial plants operated by the collectives. For example, when some plants accepted subcontracts from urban stores or factories, or when they served beyond local areas, party cadres considered their activities to be capitalist profit-making and ordered them stopped. This is discussed further in Chapter 6.

In short, the policies of the Cultural Revolution decade attempted to develop agriculture, but only in a highly ideological and restrictive way. As a result, only limited gains were made in agricultural development. In spite of the excessive emphasis on grain production, grain output stagnated in 1968–69, 1971–72 and 1975–77 (see Table 3.1). Although poor weather played a role in these periods, inappropriate policies contributed to the problems. As the population continued to grow, and reached or exceeded 950 million in 1976, China's agricultural situation became quite precarious. Changes were obviously needed to avoid a disaster.

Post-Mao Modernization Drive Since 1977

With the death of Mao in 1976 and the subsequent purge of his radical supporters, particularly the Gang of Four, a new era in China was ushered in. Under the leadership of the pragmatic Deng Xiaoping, "Four Modernizations" (agriculture, industry, science and technology, and defense) became the focus and the rallying cry of the Chinese developmental efforts. A Ten-Year Economic Development Plan (1976–85), originally drafted in 1975, was revised and adopted. A Twenty-three-Year Plan for the rest of the century was also drafted.[20] It was envisaged that, by the year 2000, China would catch up with the industrialized countries in most areas and would even surpass them in some areas.

The cornerstones of the 1976–85 modernization plan were 120 major industrial projects to be completed by 1985. They included 10 iron and steel complexes, 9 nonferrous metal complexes, 8 coal mine projects, 10 oil and gas fields, 30 power stations, 6 new trunk rail lines, and 5 harbors. The nature of these industrial projects resembled that of the First Five-Year Plan based on the Soviet model of development, that is, capital- and technology-intensive with emphasis on heavy industry. It is clear that the ranking of investment priority was heavy industry, light industry, and then agriculture. In agriculture, farm mechanization was given top priority. The production of tractors, other farm machines, and chemical fertilizer was to be greatly increased. The major objective of farm mechanization was to raise agricultural labor productivity. It was envisaged that, by 1985, grain output would be increased to 400 million metric tons from the level of 285 million metric tons in 1977, and that at least 85 percent of all major farm work would be mechanized by 1985.

In the initial euphoria over modernization, the material and financial requirements of the projects were not carefully calculated. In early 1979, growing imbalances in the economy and the rising cost of capital investment forced a more realistic reappraisal of the remaining period of the Sixth Five-Year Plan and of the Ten-Year Plan. There were two major considerations in the reappraisal.

First, in terms of investment priority, the original ranking was heavy industry, light industry, and agriculture. Agriculture had the slowest growth in the past, although China needs higher agricultural growth to feed the population and to cut down on the expensive import of wheat. Also, China can export light industrial products and processed agricultural products to earn foreign exchanges. Unlike investment in light industry and agriculture, investment in heavy industry will take a long time before it will yield profits. And even then, China still may not have enough heavy industry for its economic and military needs.

Second, within heavy industry itself, there were imbalances:

coal, electric power, petroleum, transportation, and building material industries have not developed fast enough to meet the needs of the metallurgical, machine-building, and processing industries. There was also a shortage of raw materials.[21]

As a result, adjustments were made in the economic plan in 1979. The order of investment priority was reversed, from agriculture to light industry, and then to heavy industry. Some of the 120 key investment projects would be postponed or scaled down. On the other hand, "projects that are easy and quick to build and economically profitable" would be speeded up.[22] Clearly the Chinese planners had become more realistic and cost-benefit conscious.

In agriculture, the new development strategy had led to many important policy changes. The new emphasis since 1979 has been on building a diversified and prosperous agriculture through pragmatic policies and the application of science and technology. A diversified agriculture for the nation means that forestry, animal husbandry, and fisheries, as well as farming, are to be simultaneously developed, in contrast to the previous policy of emphasizing grain production at the expense of the others. For regions with unique natural conditions, this implies greater specialization in accordance with their comparative advantage, such as sugarcane production in some southern areas and livestock production in Inner Mongolia.

The new emphasis on rural prosperity is aimed at increasing the peasants' incentives to produce and to raise productivity. During the Cultural Revolution decade, the extreme leftist ideology tended, in practice if not in theory, to regard high peasant income as incompatible with socialism. Local party cadres often discouraged or prohibited "improper" money-making activities by the collectives or the peasant households. This attitude is now attacked by the leadership as the "sham socialism of poverty," and rural areas are encouraged to "become rich" by raising agricultural yields and by developing collective enterprises and household sideline production.

To increase the work incentives and incomes of the peasants, the government reintroduced or reaffirmed the following policies:

1. The distribution of income is based on work performed, not on the principle of egalitarianism. The production team, the lowest level of the rural collectives, is the basic accounting unit. It

may make contracts with peasant households or individual peasants for production quota or for labor. If output exceeds the contract figures, the households or individuals may keep the excess as rewards. This is called the "responsibility system" in agricultural production.

2. Private plots, which were previously eliminated or collectively cultivated, have been returned to the peasant households. The size of the plots has also been increased in some areas. For example, in Sichuan where many of the economic changes were first introduced, private plots have increased from 4 percent of the cultivated land in 1962 to 7 percent in 1978, 10 percent in 1979, and 15 percent in 1980; it might be further raised to 20 percent in the future.[23] In addition to private plots, the collectives may allocate a certain amount of hill wasteland to the peasants for planting trees, herbs, or grass; these products will be kept by the peasants.

3. Rural free markets have been reopened for agricultural and sideline products. Agricultural products that are outside the state production plan or are over and above the production quotas of the collectives may be sold in the markets; the same is true with sideline products produced by the peasant households. During the Cultural Revolution decade, state commerical organizations monopolized the purchase and sales of most of these products.

4. The collectives have the "right of self-management" in implementing the above policies as well as in carrying out their normal production activities. Party cadres are instructed to respect this right and to refrain from blind interference in the collective's activities.

In implementing these policies, it is inevitable that the disparity in income between the richer collectives and households and the poorer ones will grow. This is considered to be temporary and consistent with the law of socialist development. The reason is that it will permit the rich ones with the appropriate conditions to develop at a faster pace and thus "strengthen the material basis for socialism," and help provide the investment funds for agricultural modernization, particularly farm mechanization. In addition, the experience of the richer collectives will provide models for the lagging ones to emulate and stimulate their "mass labor enthusiam through striving to catch up."[24]

To raise the incomes of the rural areas, the government raised the procurement prices for agricultural products in early 1979 (see Chapter 6). State investment in agriculture has increased from 10.7 percent of the budgeted funds for capital construction in 1978 to 14 percent in 1979.[25] The target is to increase it to 18 percent by 1985. State loans for agriculture will be more than doubled by 1985.[26]

To promote modern agricultural technology, the government introduced various changes. The agricultural machinery industry is being reorganized to achieve greater specialization, product standardization, and higher quality. More chemical fertilizer is being produced. Improved varieties of rice are being disseminated at a faster pace (see Chapters 4 and 5). Basic as well as applied agricultural research is being stepped up. Finally, technical education in the rural areas is being promoted, including the technical education of many party cadres who are ignorant of agricultural technology. The status of technicians and experts is being raised.

In short, the general philosophy of the post-Mao period is that "in China the capitalist and landlord classes and their whole system of exploitation have already been abolished, and the only source of wealth is the development of productive forces."[27] Economic incentives and better technology are held as the key to the development of productive forces in agriculture.

4
Technological Changes: Fertilization, Water Control, and Plant Breeding

The importance of technological changes to China's agriculture has been mentioned in previous chapters. Chapters 4 and 5 examine some vital areas of agricultural technology—fertilization, irrigation, better crop varieties, mechanization, rural energy, and plant protection—that have contributed to China's agricultural development.

The term "technological changes" in agriculture refers to the development, adoption, and adaptation of improved agricultural inputs and practices to increase farm output per unit of land or labor. Because technological changes take place within the specific context of a society, it is important to note three institutional characteristics of China's technological changes in agriculture.

First, the research and development of a better technology takes place simultaneously in the state sector and the collective sector, i.e., at government agricultural research institutions and at the experimental farms of the communes and production brigades. The institutions are concerned with both basic and applied research, providing leadership and technical assistance to the collective sector. The farms are concerned with applied research of local interest, such as adaptation and problem-solving. This is an example of the strategy of "walking on two legs" as applied to agricultural research.

Second, the production of modern agricultural inputs, such as chemical fertilizers, seeds, and farm machines, also has this dualistic nature: modern state enterprises and medium- and small-scale rural plants coexist to supply these inputs. The former are large and capital intensive, and often utilize imported equipment and technology. The latter are more labor intensive and use less sophisticated technology; they are operated by both the county governments and the communes and brigades as part of the rural industries.

Third, with the exception of the state farms,[1] decisions at the local level concerning the adoption of better inputs and practices are in principle made by the rural collectives themselves, i.e., by the communes, production brigades, or production teams. However, as in crop production decisions, government policies and party cadres can influence the decisions made. The government can influence the decisions by changes in the prices of the inputs the government supplies to the collectives (see Chapter 6). Local party secretaries can exert pressures and interfere in the decision making of the collectives (see Chapter 7).

Fertilization

Organic Fertilizers

As China's arable land is almost fully utilized and the scope for reclamation is limited, increases in agricultural output have to come from increases in yield per unit of land. Among the measures to increase yield, increased application of fertilizer is the most important.

Historically, the Chinese peasants have long relied on the use of organic fertilizers to maintain soil fertility. As early as the Yuan dynasty (1271–1368), night soil, lime, mud from ponds and rivers, and grasses, straw, etc., were used as fertilizers. Later, probably around 1500, bean cake was also used—an important innovation in a then technologically stagnant agriculture.[2]

Night soil and animal manure subsequently became the two major sources of organic fertilizers because of their high nutrient content, low resource cost, and availability. Other organic fertilizers, such as compost and mud, have low nutrient content. Bean cakes have high nutrient content, but they are expensive because they require valuable cropland for production. As China's population grew more rapidly than its cultivated land, the supply of night soil per unit of land increased accordingly. The supply of hog manure also increased with population growth because pork is the principal meat consumed by the Chinese and because hog feed is inexpensive.[3]

After 1949, the government promoted heavier fertilization of cropland to raise farm production, and the Chinese peasants used even more organic fertilizers than before. For example, in 1929–33, according to Buck's survey, an average of 7,472 kilograms of night soil and animal manure per hectare of cropland was used in China's major agricultural regions.[4] In 1952, Chinese sources indicate that an average of 11,250 kilograms of organic fertilizers per hectare was applied to 60 percent of cultivated land; in 1957, the figures rose to 15,000 kilograms per hectare on 80 percent of cultivated land.[5] In 1957–58, as part of the rural mobilization campaign of the Great Leap Forward, the peasants were mobilized to collect and utilize all sorts of organic fertilizers. As a result, there was some increase in the amount of organic fertilizers used, as can be seen in Table 4.1. However, as the sources of better organic fertilizers were quickly exhausted, the peasants resorted to collecting and using inferior fertilizers, such as river and pond mud. Even rubbish and industrial wastes, which had dubious nutrient value, were collected. There was much waste of labor, which no doubt was the result of the government's exhortation that "manure can be found everywhere and fertilizers can be made from almost anything."[6]

In the early 1960s, the emphasis was shifted to the use of chemical fertilizers. Nevertheless, the use of traditional organic fertilizers continued to increase, although at a slower rate than that of chemical fertilizers. In terms of plant nutrients, the amount of chemical fertilizers used increased 13 times during 1960–77 while that of organic fertilizers increased 1.8 times during the same period. The ratio of nutrients between organic and chemical fertilizers was about 20 to 1 in 1960 and 2.7 to 1 in 1977.[7] Thus, China continues to use much more organic than chemical fertilizers, although the latter have been catching up very rapidly.

Organic fertilizers will continue to dominate over chemical fertilizers in China's agriculture. At the present level of industrialization and economic development in China, the chemical fertilizer industry is simply incapable of supplying all the fertilizers needed

TABLE 4.1
Organic Fertilizers: Major Sources and Nutrient Content, 1952-77
(million metric tons)

Year	Night Soil	Hog Manure	Draft Animal Manure	Total Nutrient Content[a]
1952	186	130	422	10.1
1955	216	131	511	11.7
1957	237	184	514	13.0
1958	241	256	524	14.2
1960	250	169	490	12.5
1962	257	185	437	12.2
1965	311	302	632	17.0
1967	324	358	599	17.6
1970	346	379	609	18.4
1972	362	468	659	20.6
1975	385	483	720	22.1
1976	392	490	741	22.5
1977	398	492	763	23.0

Source: Anthony M. Tang & Bruce Stone, Food Production in the People's Republic of China (Washington, D.C.: International Food Policy Research Institute, 1980), p. 61.

[a]Includes green manure, oil cake, compost, mud as well as night soil, hog manure, and draft animal manure. It is calculated as the sum of the components, each weighted by its ratio of plant nutrient to gross weight. See ibid., p. 64.

for agricultural production. To do so would have required much more resources than China has available. Nor does China have the foreign exchanges needed to import large quantities of chemical fertilizers. However, as China's human and animal populations continue to grow, night soil and animal manure will necessarily increase in supply. Because these and other organic fertilizers are produced locally by peasant households and rural collectives, the peasants and collectives have a high degree of self-reliance in the fertilization of their cropland. In addition, organic fertilizers have very low resource cost because no foreign exchange is needed and only locally available materials are used.

For these reasons, during the Cultural Revolution, the raising of hogs, particularly by the collectives, was promoted by the party leadership as a self-reliant way for the peasants to increase both meat consumption and fertilizer supply. Mao Zedong's saying that "each pig is a small fertilizer factory" was much publicized. Both the peasant households and the collectives are permitted to raise hogs. The households can use the manure on their private plots or sell it to the collectives. Consequently, there was a substantial increase in the supply of hog manure between 1965 and 1975, as shown in Table 4.1.

Since the mid-1970s, the utilization of night soil and animal manure has been improved in parts of the country with the widespread construction of fermentation tanks for generating biogas. In 1980, there are more than seven million such tanks throughout China's countryside (see Chapter 5). Night soil, animal manure, and organic waste, such as crop stalks, are stored in the tanks for fermentation. Not only does this generate methane gas for family cooking and lighting, but it also gives the organic fertilizers 16 percent more nitrogen and 25 percent more phosphorus.

After night soil and animal manure, the next important source of organic fertilizer is green manure, which refers to cover crops that are plowed under while green. In South China, winter green manure, such as milk vetch, cow vetch, and summer sesbania are grown. North China grows alfalfa and hairy vetch. South China grows more green manure crops because of its longer growing season. In recent years, it has become increasingly popular in the south to grow red duckweed in rice fields and other water surfaces as a source of manure and fodder.

Since the early 1960s, the peasants have been encouraged to grow green manure crops on fallow land. In 1964, green manure crops were grown on more than 5 million hectares of land.[8] By the mid-1970s, the south alone had more than 9 million hectares of winter green-manure crops. Since then, however, the amount of land planted to them has been declining gradually, the major reason being the fear on the part of the peasants and party cadres that green manure crops may compete with grain and cotton and adversely affect their production. Nevertheless, as green manure has high and balanced nutrient content, Chinese scientists have concluded that it is important to encourage the peasants to use it.[9]

Chemical Fertilizers

The government realizes that organic fertilizers are necessarily limited in supply; therefore, it has increasingly paid more atten-

TABLE 4.2
Chemical Fertilizers: Production and Imports, 1952-80
(million metric tons of nutrients)

Year	Production	Imports	Year	Production	Imports
1952	0.04	0.04	1972	3.58	1.35
1957	0.16	0.27	1973	4.23	1.59
1961	0.36	0.23	1974	4.69	1.20
1965	1.48	0.64	1975	4.72	1.36
1967	1.76	0.86	1976	5.80	0.77
1968	1.95	1.42	1977	7.24	1.48
1969	2.16	1.53	1978	8.69	1.71
1970	2.45	1.72	1979	10.65	1.73
1971	3.05	1.37	1980	12.32	n.a.

Sources: Joint Economic Committee, Chinese Economy Post-Mao, p. 650 for 1952-65; FAO Fertilizer Yearbook, 1978 (Rome: FAO, 1979) for 1967-75; FAO Fertilizer Yearbook, 1980 for 1976-79; Beijing Review, No. 19, May 11, 1981, p. 24 for 1980 production figure.

n.a. = not available.

tion to the production of chemical fertilizers, especially since the early 1960s. Domestic production has risen rapidly, from less than 1 million metric tons of nutrients per year in the early 1960s to more than 2 million metric tons in the late 1960s, 8.7 million metric tons in 1978, and to 12.3 million metric tons in 1980 (see Table 4.2). Imports of chemical fertilizers also increased throughout the 1960s, but remained at an average annual rate of 1.5 million metric tons of nutrients in the 1970s.

With increases in domestic production and imports, the application of chemical fertilizers has also increased rapidly. It has risen from an annual average of 13.3 kilograms of nutrients per hectare of cropland in 1961–65 to 45.5 kilograms in 1972, 74.3 kilograms in 1977, and 109.2 kilograms in 1979. Compared with the world average of 27.9 kilograms per hectare of cropland in 1961–65, 54.3 kilograms in 1972, 68 kilograms in 1977, and 77.1 kilograms in 1979, the consumption of chemical fertilizers has grown more rapidly in China than in the rest of the world (see Table 4.3). However, China's consumption of fertilizer per hectare of cropland is only at an intermediate level compared with that of other Asian countries. It is higher than that of the South

TABLE 4.3
International Comparison of Consumption of Chemical Fertilizers Per Hectare of Crop Area, 1961-79
(kilograms of nutrients)

Country	1961-65[a]	1972	1977	1979
China	13.3[b]	45.5[b]	74.3[b]	109.2
Bangladesh	4.4	20.0	37.1	44.6
Burma	0.7	4.6	6.1	10.5
India	3.7	16.7	25.3	29.6
Indonesia	8.4	28.9	35.0	44.1
Japan	305.2	389.5	428.1	477.7
Korea, North	78.3	176.2	276.3	336.0
Korea, South	157.0	288.9	329.9	383.6
Malaysia	9.4	35.5	49.6	103.2
Pakistan	3.4	22.8	35.1	51.9
Philippines	13.3	25.6	32.2	34.6
Taiwan	201.8[c]	150.3[c]	282.6[c]	261.1[c]
Thailand	2.2	10.8	15.6	17.4
Asian Average	11.8	31.0	45.4	61.5
World Average	27.9	54.3	68.0	77.1

Sources: FAO Fertilizer Yearbook, 1978 and FAO Fertilizer Yearbook, 1980 for all countries except Taiwan. Annual Economic Report of China, 1981 (Beijing: Academy of Social Sciences, 1981), part VI, p. 13 for China's 1979 figure. Figures for Taiwan are calculated from Taiwan Statistical Data Book, 1981, pp. 67-68.

[a]Annual average.
[b]FAO estimates.
[c]Weighted average of rice and sugarcane crops only.

and Southeast Asian countries, but much lower than that of the other East Asian countries, as shown in Table 4.3.

In the production of chemical fertilizers, China has adopted the "walking on two legs" strategy of simultaneously developing large and small plants. Large plants are often built with imported equipment and technical assistance from the Western countries and Japan. The small plants utilize Chinese-made equipment; the production process is more labor-intensive and the production capacity is less than 15,000 metric tons in gross weight.[10] The small plants are less efficient, but they require less initial investment, and because they are located in the rural areas, the cost of distributing the product is lower. They are also carefully designed and constructed, unlike the small plants that were hastily erected during the Great Leap Forward. In 1964, the small plants produced 30 percent of the total chemical fertilizers produced in China. The share of the small plants increased to 60 percent in 1970–72, but declined to 45 percent in 1974 with the expansion of the modern plants.[11] In 1979, China had 1,533 small nitrogen fertilizer plants, which produced 55 percent of the nation's output of nitrogen fertilizer, or 45 percent of the total output of chemical fertilizers.[12]

Chemical fertilizers produced by the state enterprises are sold by the supply and marketing cooperatives to the rural collectives.[13] The prices charged for them have been successively reduced (see Chapter 6). This reflects in part the government's aid to agriculture and in part the reduction in the cost of production as more modern plants are built. The distribution of chemical fertilizers is used by the government as an instrument of the state's policies on crop production and procurement. The favored crops for fertilizer distribution have been cotton, rice, and wheat, with cotton being usually accorded the highest priority. However, some peasants reportedly have used fertilizers allocated for cotton on other crops.[14] The collectives receiving fertilizers are required to fulfill certain quotas for cotton or grain procurement. Thus the objectives of fertilizer distribution policies are not only to increase production, but also to affect rural consumption, savings, and investment.[15]

In addition, the government is attempting to maximize the benefits of fertilizer application by favoring certain areas in distribution. Generally, areas with high yields in priority crops and with better irrigation facilities have been favored.[16] This preferential policy for the distribution of chemical fertilizers has contributed to the widespread adoption of fertilizer-responsive, high-yielding varieties of rice in the south since the 1960s. Because fertilizer application has to be accompanied by water, and because North China is not as well irrigated as the south, the level of fertilizer application is lower in the north.

But there are problems of waste and inefficiency in the distribution and application of chemical fertilizers. First, due to poor packaging, transporting, and storage, the spoilage rate is very high. It has been estimated that as much as 20 percent of the bags are damaged, causing an annual loss of up to 1.2 million tons of fertilizers, which is more than the annual output of a large synthetic ammonium plant.[17]

A second problem stems from the imbalance between the nutrient needs of the soils and the types of fertilizers applied. There are two reasons for this imbalance: (1) lack of detailed information about soil nutrients, and (2) the imbalance in fertilizer production. Currently China lacks detailed information on the soil nutrients of most of its farming regions. Although the first national soil survey was conducted in 1958–60, the quality of the survey was not high. Conducted during the Great Leap Forward, the survey was concerned with the preliminary classification of the soils and with summing up the peasants' experiences in land utilization and soil improvement.[18] It did not gather comprehensive and detailed information on the soil conditions of local farming areas. On the basis of the survey, a study in 1962 estimated that 80 percent of the soils in China were deficient in nitrogen, 50 percent in phosphorus, and 15 percent in potassium.[19] However, a more recent estimate suggests that two-thirds of the cultivated land in the country are deficient in phosphorus and one-third in potassium.[20] These differing national figures offer no practical guide to the peasants. A second national soil survey, started in early 1979, is yet to be completed. At the end of 1980, only 142 counties in a pilot project have been surveyed.[21] One of the major findings of the pilot project, not surprisingly, is that most farming communities do not know the exact soil conditions of their land and have not applied the appropriate mix of nutrients.

Even where the nutrient requirements are known, the appropriate fertilizers are often unavailable. China has so far concentrated on the production of nitrogen fertilizers, such as ammonium nitrate and urea. The production of phosphorus is much less, while that of potassium fertilizer is insignificant. For example, in 1979, the gross output ratio among nitrogen, phosphorus, and potassium fertilizers was 1:0.2:0.04 whereas in industrialized countries it is generally about 1:0.7:0.6.[22] The government distributes fertilizers to the rural collectives, not according to what they need or might need, but according to what is available. Consequently, there have been complaints that much farmland in the south is deficient in potassium but that the needed fertilizer is not available. Other agricultural regions also suffer from unbalanced application of chemical fertilizers because of the unavailability of the appropriate fertilizers.[23]

Water Control and Irrigation

Like other ancient civilizations, China has a long history of water control. Flood-control schemes were constructed centuries before the empire was unified in 221 B.C., and the oldest irrigation system, still in existence, was built around 200 B.C. on the Chengdu Plain in Sichuan province.[24] Throughout the dynasties, more than 50,000 water-control projects were recorded in various local historical gazettes. The bulk of them were constructed after the tenth century, during periods of new land settlement and dynastic growth.[25]

Water-control projects in China are for the purposes of flood control and irrigation. Both are essential because large areas of China are subject to severe floods and/or droughts. This stems from the fact, discussed in Chapter 1, that water resources in China are very unevenly distributed both regionally and seasonally. The Changjiang Basin and areas south of it have 75 percent of the country's total surface runoff; the basins of the Huanghe, Huaihe, and Haihe in the north have only 3.8 percent.[26] Furthermore, the rainfall is concentrated in the summer, making it necessary to control flooding in some river basins during the summer and to irrigate some farmland during the dry seasons. The north does not have enough surface water for crop irrigation, but large areas along the Huanghe in the river basins are subject to severe flooding because the Huanghe is heavily silted.

From the very beginning of the People's Republic, the Chinese leadership has recognized the great importance of water control and has paid close attention to it. In his "Eight-Character Charter of Agriculture," Mao Zedong singled out water as the most important factor for China's agriculture.[27] Before 1957, the emphasis of China's water-control efforts was on flood control because floods were traditionally considered by the Chinese to be more dangerous than droughts.[28] As the Huanghe, Changjiang, Huaihe, and Haihe are the major sources of flooding, most projects were centered on them. These projects included the reinforcement of dikes, the construction of large multipurpose reservoirs for flood control and electricity generation, and the diversion of the river water. For the Huanghe, the most treacherous river in China, a complex system of reservoirs was built in the upper reaches. A large afforestation program was launched to decrease soil erosion and silting of the river. In the lower reaches, spillways and reservoirs were also constructed. The largest reservoir, Sanmen Gorge in Henan province, was originally designed with Soviet assistance in 1955. Started in 1957, it was completed in 1960 but had to be rebuilt because the silt accumulation was much worse than expected. On the Changjiang, the Qing River flood-control project was built in 1952 between Hubei and Hunan, an area along its middle reaches where floods had been most serious.[29]

Although flood-control activities were continued throughout the 1960s and 1970s, the emphasis of water conservancy was shifted after 1957 to irrigation. Prior to 1957, the irrigated area was expanded primarily through the restoration and improvement of the old irrigation systems. In 1957–58, as part of the Great Leap Forward to mobilize rural resources for agricultural development, the leadership emphasized the construction of a large number of small irrigation projects, such as canals, ditches, and ponds. Unlike the large flood-control reservoirs built along the major rivers, these small irrigation projects were the responsibility of the local authorities and the newly established rural collectives.

These local units lacked the expertise and coordination needed in the construction of irrigation systems, but tremendous political pressure was exerted to mobilize rural labor to construct as many projects as possible. As a result, many projects with poor design and low quality were hastily constructed by inexperienced peasants. River dikes were recklessly cut to build water-diverting channels, and many channels did not have proper drainage.[30] Total farmland also declined because of excessive construction. Heavy rains in 1959 caused one of the worst floods in China's history, lasting for three consecutive years. Most of the new water works were destroyed; the irrigated area declined from 38.2 million hectares in 1957 to 32.1 million hectares in 1961; gradual salinization of farmland resulted; and agricultural production declined. It was not until 1967 that irrigated land was restored to the 1957 level (see Table 4.4).

TABLE 4.4
Cultivated Area and Irrigated Area, 1949-80
(million hectares)

Year	Cultivated Area[a]	Irrigated Area	Percentage of Cultivated Area Irrigated
1949	97.9	26.1	26.7%
1952	107.9	30.7	28.5
1957	111.8	38.2	34.2
1961	107.1	32.1	30.0
1965	103.6	35.1	33.9
1967	107.0	38.0	35.5
1975	99.7	46.2	46.3
1978	99.5[b]	46.7	46.8
1979	99.5	47.7	47.9
1980	99.5[b]	47.3	47.5

Sources: Annual Economic Report of China, 1981, part VI, p. 9 for all years except 1961, 1967, 1978, and 1980; Tang & Stone, Food Production in the People's Republic of China, p. 51 for 1961 and 1967; U.S. Department of Agriculture, Agricultural Situation: Review of 1980 and Outlook for 1981 (People's Republic of China), 1981, p. 27 for 1978 and 1980.

[a]China's State Statistical Bureau indicates that its figures for cultivated area are underestimates.
[b]Adjusted upward from 99.0 to 99.5 by the author in the light of China's figure for 1979.

In the early 1960s, the emphasis was shifted to repairing or improving existing irrigation facilities, including the addition of drainage outlets. North China was the focus of these efforts because it was most severely affected by the floods of the late 1950s. In the mid-1960s, a large number of pumps were produced and installed in the tube wells drilled in North China. Since then, more than one million tube wells have been drilled, which irrigate about seven million hectares of land in the north.[31]

From 1965 to 1975, with continuous improvement in irrigation and drainage, the irrigated area has been expanded at an average rate of slightly over one million hectares per year. In 1980, total irrigated area was 47.3 million hectares, about 48 percent of total cultivated land, compared with less than 10 percent in the United States. The expansion in irrigated area has made it possible to extend multiple cropping and to achieve stable and high yields in many areas.

The north remains more deficient in irrigation than the south. To alleviate this deficiency, three measures can be undertaken. First, the Chinese can continue the drilling of tube wells, equipped with power pumps. One potential danger in this approach is that the groundwater level may be lowered with massive pumping, making further utilization of this water increasingly difficult. In fact, some areas in Hebei province have already experienced that since the late 1970s because of the excessive drilling of tube wells, and 228,000 of the 488,000 tube wells drilled there during 1973–78 can no longer be used.[32]

Second, the Chinese recently have been actively considering large-scale diversion of water from the Changjiang to areas north of the Huanghe. The project will involve building dozens of pumping stations along each of three proposed routes. There is still no consensus among the Chinese scientists on the desirability of the project: Many are in favor if it, but some have various reservations. One concern is that the project many cause a water shortage along the middle reaches of the Changjiang when multiple cropping of rice, which is heavily dependent on irrigation, is expanded there.[33] Another argument is that there exists sufficient groundwater in North China and that it is better to utilize it more efficiently than to divert water from the south.[34] One final concern is that water diversion may change the ecosystem and bring about the salinization of soils.[35]

Third, the removal of silt from the Huanghe can make more river water usable for irrigation without clogging irrigation canals. In the long run, this has to be coupled with the cessation of soil erosion, particularly in the Loess Plateau, to prevent further silting of the river. This will involve massive reforestation. A long history of deforestation coupled with excessive conversion of wooded areas and grasslands into farmland since the mid-1960s have resulted in increased soil erosion in the north and the northwest and increased silting of the Huanghe. As part of the efforts to reverse this, the government adopted in early 1979 a new reforestation program to plant 27 million hectares of trees in the following seven years. Large tracts of grassland are also being restored to help fight soil erosion.

Arable land is extremely limited and reservoirs and irrigation channels take up much land, so the Chinese have introduced in recent years some land-saving irrigation methods and projects. Spray irrigation, introduced in many areas, economizes on the use of water, labor, and land. It is particularly useful in mountainous areas where alternative methods of irrigation are more difficult. Thus, it has become one of the priorities of the current agricultural modernization program.

Some rural collectives have constructed underground irrigation channels to replace those above ground. There are several advantages to this approach: lower pumping cost, fewer maintenance problems, the elimination of evaporation, and greater flexibility in channel routing.[36] The disadvantage is that the initial construction cost is much higher than above-ground irrigation. It is likely that this method of land-saving irrigation will be increasingly adopted in densely populated farming areas.

The first underground reservoir in China was recently constructed. Located in Nangong county, Hebei province, it was completed in early 1980. This reservoir was made possible by the discovery of a large ancient river channel buried underground. It has a storage capacity of 480 million cubic meters and offers the advantages of being earthquake-proof and seepage-proof, having no evaporation, and having low maintenance cost. According to preliminary findings, there are more than 20 sites in Hebei Plain alone that are suitable for the construction of similar underground reservoirs.[37]

Plant Breeding

The development of improved crop varieties by scientific methods had a slow start in China. Although Chinese farmers in the past had developed a great number of crop varieties, the process was slow and was based on the field experience of farmers who selected those varieties best suited to local conditions. From the beginning of the People's Republic, the government attempted to encourage and accelerate the development of better seeds. However, in the early 1950s, China still relied primarily on local peasants to select and popularize better seeds themselves. Although the government encouraged exchanges among different areas, the benefits of the seed-improvement program were very limited. It was not until the mid-1950s that systematic breeding by agricultural research organizations became important. The Great Leap Forward of the late 1950s further expanded the scale of the program nationwide.

Rice

In 1956, Chinese scientists initiated the breeding program that led to the first high-yielding dwarf indica varieties of rice developed in China. These varieties had high yield potential, were responsive to fertilizers, and were relatively resistant to lodging and diseases.[38] With adequate fertilizer and water, they produced yields of 5–6 metric tons per hectare, comparable to those of the IR-8 dwarf rice developed in the Philippines. The Chinese varieties, however, had a shorter growing period of 110–115 days, making it possible to expand double cropping of rice in the south.[39]

In 1964, the Chinese began to distribute the new varieties for large-scale adoption and their use spread rapidly. By 1965, 3.3 million hectares in the south were sown to these varieties, and by 1977, these varieties were grown on 80 percent of total rice land in China.[40] This rapid diffusion of improved varieties was facilitated by the rural collective system with its "four-level research network" (county, commune, brigade, and team). The process of seed selection for local adaptability was shortened because experiments could be simultaneously undertaken at various levels.

In the late 1960s, research in rice breeding was disrupted and reduced in scale by the political upheavals of the Cultural Revolution. Thus, no further success came until 1973 when a new hybrid rice was developed. It was a cross between a wild rice from Hainan Island as the female parent and a variety from abroad as the male parent. The female parent carried male-sterility genes but the progeny, however, are fertile and can self-pollinate to produce seeds.[41] This hybrid rice has various merits: increased yield, about 20 percent or from 750 to 1,500 kilograms more per hectare than regular rice; strong root system; and better tolerance of alkalinity, drought, and waterlogging. In addition, it does not lodge easily and contains more protein. Its shortcomings are a longer maturity period, less resistance to diseases, and greater susceptibility to temperature changes.[42] Thus more field work is needed for local adaptation. In 1976, hybrid seeds were distributed for farm production on 130,000 hectares of land, which expanded to 2 million hectares in 1977.[43] In 1979, more than 5 million hectares of land were planted to this hybrid rice. The increase in yields from these plants accounted for one-third of the total increase in rice production throughout China in 1979.[44]

Wheat

The first success with breeding of wheat was the development of high-yielding, tall winter-wheat varieties in the early 1960s. In the early 1970s, the first high-yielding, dwarf winter-wheat varieties were developed.[45] Some of these were crosses between Korean dwarf winter wheat and tall superior Chinese varieties. Later varieties were Chinese-Mexican crosses, which were disease-resistant. Since then they have become the main winter-wheat varieties south of the Changjiang and have been planted as spring wheat in the northeast.[46] Adoption in North China has also spread, due to the expansion of tube-well irrigation.

A discovery of great value was that made by a peasant woman of a male-sterile wheat plant in 1972 in Shanxi province.[47] As with rice, wheat is a monoecious plant, having male and female organs in the same individual. To avoid self-pollination, it is necessary, but very time-consuming, to remove the stamens before crossbreeding. With male-sterile plants controlled by a dominant monogene, crossbreeding can be done in a much shorter time. Thus, within a few years after the discovery, the male-sterile plant has been crossbred with different wheat varieties and has produced generations of good-quality, high- and stable-yielding new wheat strains.[48] Experiments with these varieties and others are continuing on a large scale in various parts of the country.

Other Crops

Beside rice and wheat, varietal improvement has also been made in other crops, including corn, sorghum, and cotton. Hybrid corn and sorghum have been popularized since the late 1960s and have contributed to increasing the crop yields. There are also improved varieties of millet, soybeans, barley, and many fruit and vegetable crops.[49] Improved varieties of cotton were selected for popularization in 1962.[50] In the 1970s, a good cotton variety imported from the United States was found to be highly adaptable to local conditions and has become the main strain in the cotton belts along the Huanghe and Huaihe rivers.[51] It has also been crossbred with local strains to produce improved varieties. The most important development came in 1976 when a new high-yielding strain of cotton, called Lumian No. 1, was developed by the Cotton Research Institute of Shandong Province. In 1979 the strain was planted in Shandong on a large scale for demonstration purposes. Because of its superior qualities—high yield, pest resistance, and early maturity—within two years 80 percent of the cotton fields in Shandong was planted to this variety.[52]

Seed Multiplication

One important aspect of plant breeding concerns the multiplication of improved seeds. During the 1970s, decentralized seed propagation and breeding networks were set up in China's main farming areas, involving seed units at four levels: county, commune, brigade, and team. The county seed farms purify and regenerate the existing fine seed strains, test promising seeds, and also breed new strains. The commune seed-breeding stations are also involved in seed selection, testing, and breeding. Pure seeds are then supplied by the communes to the brigades and teams for their seed plots for multiplication.[53] The county will supply additional seeds when needed. For rapid multiplication of improved varieties, the harvested seeds are immediately sown as the second crop—at the same station in the case of rice, or in a warmer region in the case of cotton.[54]

The advantages of this decentralized seed-breeding system are (1) it enables the collectives to be self-reliant in the supply of seeds, (2) it reduces the possibility of diseases spread by seeds from a common seed source,[55] and (3) it promotes rapid popularization of new varieties. New varieties are tested both at the county seed farms and on the experimental plots of the communes and brigades. Within a minimum of time, the characteristics of the new varieties can be obtained and evaluated for local selection.

The weakness of the system is that the collectives tend to have poor facilities and a low level of technical sophistication. As a result, the multiplication of seeds has been plagued with the problems of mixing with inferior strains and of degeneration. The saying "pure in the first year, mixed up in the second year, and degeneration in the third year" is indicative of the problems.[56] Consequently, the high yields initially achieved often could not be maintained. Further, in the case of certain virus diseases, it is unlikely that the communes have the technical sophistication to detect the viruses and select virus-free materials for seed or for propagation.[57]

To eliminate these problems, the government since 1978 has stressed the importance of standardization, quality control, and specialization in the production of seeds. Specialized seed-multiplication units outside of the communes have become increasingly important. Seed companies at the county level have been expanded. The China Seed Company and its regional branches have been set up to facilitate imports of seeds from abroad and to develop modern seed production. However, in 1980, the seed companies at various levels supplied only 10 percent of the seeds used for all crops.[58] As of late 1981, only some 300 counties out of more than 2,000 in the country were engaged in the standardization and specialization in seed production at the county level.[59] Thus it will take some time before the new policy is fully implemented.

5
Technological Changes: Mechanization, Rural Energy, and Plant Protection

Agricultural Mechanization

Evolution of Policy

As early as 1959, Mao Zedong declared that "the fundamental way out for agriculture lies in mechanization." However, not until the 1970s did agricultural mechanization increasingly assume a central role in China's attempt to modernize its agriculture. In the early 1950s, the overriding issue in agricultural mechanization was whether mechanization should precede the collectivization of agriculture or vice versa, an issue on which Mao Zedong and Liu Shaoqi held opposing views. Liu was of the opinion that mechanization should precede collectivization; large quantities of tractors and other modern agricultural inputs must be given to the peasants as a tangible benefit of collectivization in order to entice them to join the collectives voluntarily. Mao's view was that the peasants must be collectivized first; only then could the requisite labor and capital be accumulated to mechanize agriculture. As for the peasants' voluntary participation in the collectives, the proper way to ensure it was through the political education of the peasants.[1] Mao's view eventually prevailed, and as discussed in Chapter 3, China's agriculture was rapidly collectivized through several stages by 1957.

Portions of the first section of this chapter have been published in Robert C. Hsu, "Agricultural Mechanization in China: Policies, Problems, and Prospects," *Asian Survey*, Vol. 19, No. 5 (May 1979), pp. 436–449.

Both Mao and Liu were influenced by the Soviet experience of collectivization and believed that big tractors were a symbol of socialism and modern agriculture. Also, both of them held to the Marxian labor theory of value in which labor was the only productive factor of production, and thus they did not distinguish between the rise in labor productivity and the rise in total output. They failed to realize that mechanization would save labor, which was abundant in China, but not arable land, which was in short supply, and that although labor productivity would be raised, total output might not increase significantly in the absence of other new agricultural inputs, such as chemical fertilizer and better seeds.[2]

In the mid-1950s, state economic planner Bo Yibo challenged both Liu's and Mao's views and contended that, in labor-surplus China, mechanization was not practical and that measures to increase crop yields per unit of land were more appropriate. Bo's analysis was supported by the results of surveys conducted in 1956 on the potential effect of agricultural mechanization.[3] As a result, the emphasis in agricultural mechanization was shifted to the transitional step of improving traditional farm implements. However, the farm implement improvement program was not completely successful. Because of inadequate research and development, the much-publicized double-wheel, double-share plows were unsuitable for use in the south.[4] On the other hand, improved traditional plows were well received by the peasants.

In the late 1950s, the Great Leap Forward diverted much farm labor to the construction of water conservation projects and backyard furnaces and created widespread labor shortages on the farms. At Mao's insistence, a ten-year program was adopted in 1959 to accelerate agricultural mechanization.[5] The Ministry of Agricultural-Machine Building Industry was established in 1959 to implement the plan. Some 600–700 farm machinery research bureaus at the county level were established all over the country.

The earlier debate between Mao and Liu on mechanization versus collectivization also led to a prolonged struggle between them over the issue of the ownership and management of agricultural machines. Mao maintained that the collectives could run agricultural mechanization faster and better than the state-controlled tractor stations; therefore, the collectives should buy and operate the tractors themselves on the basis of self-reliance—another reason why collectivization should precede mechanization. In contrast, Liu argued that the collectives did not have the resources to buy, nor the expertise to operate, the tractors; therefore, the state tractor stations should be responsible for the farm machines.

Mao's position was bolstered by a 1956 report by Kang Sheng on Soviet collective farming and on the Chinese tractor stations that were modeled after the Soviet system during the First Five-Year Plan. The report concluded that the Soviet tractor stations were contrary to the interests of the peasants and that the Chinese stations had also introduced contradictions between the peasants and the state.[6] As a result, at a 1958 national conference it was decided to transfer tractors from the tractor stations to the newly established communes.

Because of the failure of the Great Leap Forward and the decline of Mao's influence after 1960, agricultural mechanization lost its priority. As part of the move to centralize the economy and restore order, many tractors were transferred from the communes to the tractor stations. At the same time, a realistic policy of "simultaneous development of mechanization and semimechanization with the main stress on the latter" was adopted.[7]

Agricultural mechanization was further handicapped by the shortage of steel, electric power, coal, and gasoline. Problems of organization and management of the state tractor stations also plagued the program. The emphasis of the agricultural machinery industry in the early 1960s was increasingly on mechanized pumps, partly because pumps require much less steel, capital, and fuel than tractors, and partly because mechanization of irrigation and drainage was considered more urgent than that of land cultivation.

In 1964–65, the emphasis in mechanization began to be shifted to small hand tractors and other light machines, such as rice transplanters, for paddy fields. Hand tractors had been previously rejected because their plowing depth (12.7 cm) was less than that of the improved paddy plow (15.2 cm). With the increased importance of multiple cropping of rice in the mid-1960s, the advantage of greater speed of the hand tractors to relieve the

TABLE 5.1
Production of Agricultural Tractors, 1958-80

Year	Total	Conventional Tractors	Hand Tractors
In Standard Units[a]			
1958	1,100	1,100	-
1959	9,400	9,400	-
1965	23,875	23,000	875
1966	34,625	32,000	2,625
1970	79,000	70,000	9,000
1972	136,000	115,000	21,000
1975	180,000	140,000	40,000
1976	190,925	128,800	62,125
1977	241,000	156,000	85,000
In Actual Number			
1977		99,300	320,500
1978		113,500	324,200
1979		126,000	318,000
1980		98,000	218,000

Sources: Joint Economic Committee, Chinese Economy Post-Mao, pp. 292 & 318 for figures in standard units; Beijing Review, No. 27, July 6, 1979, p. 37; No. 19, May 12, 1980, p. 13 & No. 19, May 11, 1981, p. 24 for actual number.

[a]A standard unit is 15 horsepower.

labor constraint created by multiple cropping began to outweigh the disadvantage of its shallower plowing depth.[8]

During the Cultural Revolution, agricultural mechanization regained its importance. Tractors were also transferred to the communes in refutation of Liu Shaoqi's "revisionist" policy. Hand tractors became the focus of the mechanization program. With the help of the growing rural industries, the production of hand tractors increased 71 times between 1965 and 1976 (see Table 5.1). This reflected China's new strategy for agricultural development—to extend multiple cropping as much as possible to raise output, shifting from double cropping (e.g., rice-wheat) to triple cropping (e.g., rice-rice-wheat) in the south, and from single cropping to double cropping wherever possible. This strategy was made possible by various technological developments, such as improved varieties with shorter growing seasons, increased fertilizer supplies, and improved water control. Triple cropping has created a serious labor shortage during harvesting seasons when one crop has to be harvested and the next crop transplanted within a few days. Hand tractors, designed for use in paddy fields, are important in relieving the labor shortage (Figure 5.1).

Rice transplanters (Figure 5.2) have also become important for the same reason: they not only ease the back-breaking work of rice transplanting but also speed up the process considerably. Both semiautomatic and engine-propelled versions have been developed. The latter is reported to be 12 times as efficient as manual transplanting. Mass production of rice transplanters was started in 1967.[9]

An interesting innovation in the design of tractors to suit the conditions of muddy fields is the introduction of the tractor-boat (Figure 5.3), which has a flat boat-bottom to reduce resistance in the mud. Initially trial-produced around the mid-1970s in Hubei, it has become popular in some southern provinces.[10]

Post-Mao Changes

From 1977 to 1979, with the new drive toward Four Modernizations under the leadership of Deng Xiaoping, agricultural mechanization was given high priority as the basis of agricultural modernization. An ambitious agricultural mechanization program was adopted in 1978 that aimed at 70 percent mechanization by 1980 and 85 percent mechanization by 1985 of the "main work" in agriculture, forestry, animal husbandry, sideline production, and fisheries.[11] Other targets were also adopted for the production of machines and equipment for farming, irrigation, transport, and agricultural processing. Thus the scope of China's agricultural mechanization has been broadened greatly to meet modern needs.

As part of the effort to implement the new mechanization program, organizational changes have been introduced. First, greater specialization and standardization in the production of farm machines was initiated. During the Cultural Revolution decade, the stress on "self-reliance" led to a proliferation, at various levels, of small farm-machinery plants, which attempted to be self-sufficient in their production. As a result, there was also

Figure 5.1. Dongfeng-12 hand tractor. A popular model manufactured in Wuxi County, Jiangsu Province. (From the manufacturer's advertisement in *China Reconstructs*)

Figure 5.2 Automatic rice transplanter. Oil painting by Tang Mu-li, Shanghai, entitled "Spring Rain." (Reproduced in *China Reconstructs,* May 1975)

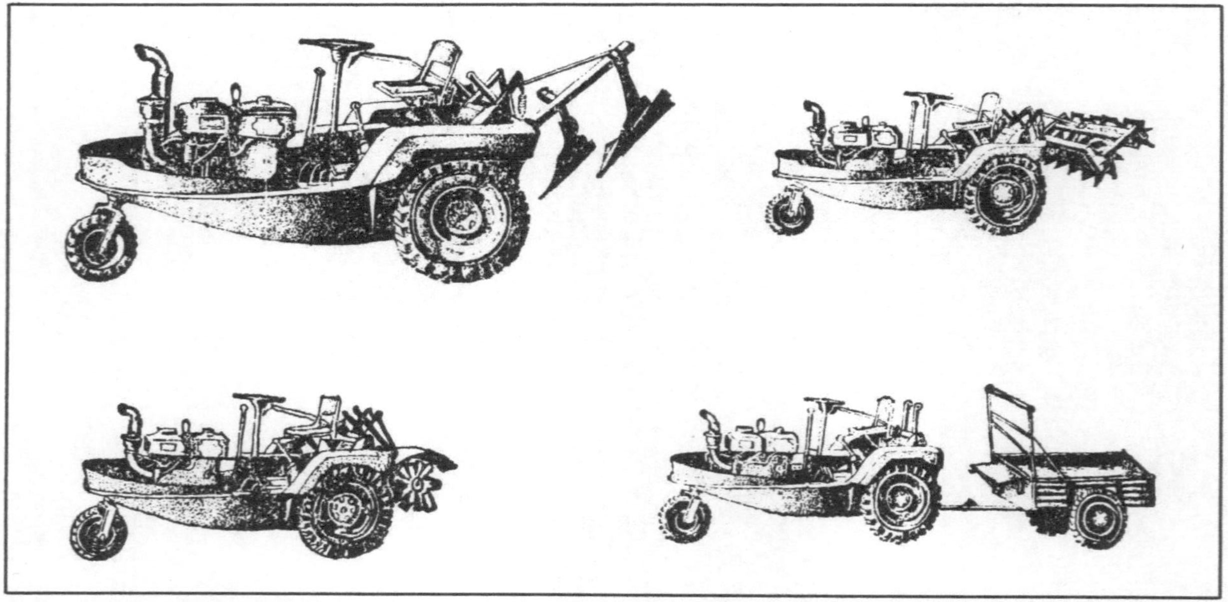

Figure 5.3. Hubei-12A Tractor-Boat. Attachments increase the versatility of the machine. (From the manufacturer's advertisement in *Renmin Ribao*)

a proliferation of models and designs without regard to standardization. The low degree of specialization contributed to the high cost of production and the low quality of products.[12] To rectify these problems, the government is encouraging the industry to raise the degree of specialization and standardization.

Second, state tractor stations were restored and the Agricultural Mechanization Service Corporation was established in 1979. The state tractor stations had been abolished during the Cultural Revolution when the ownership of tractors was transferred to the communes and brigades. It proved to be a financial burden on many collectives to own and maintain large farm machines. Consequently, as early as 1975, state tractor stations were restored in some areas.[13] In 1979, the government adopted a new policy of dual ownership. Collective ownership of farm machines will remain the dominant form of ownership, and the government will continue to provide loans to collectives with inadequate funds to purchase farm machines. At the same time, the government will set up tractor stations and rent out tractors for "reasonable fees."[14] The Agricultural Mechanization Service Corporation coordinates these stations and is managed by the Deputy Minister of Agricultural Machine Building.[15] It is in charge of the supply, maintenance, and renting of farm machinery, machine parts, and farm chemicals.

Third, manpower training programs have been organized. A high level of mechanization will require a large number of machine operators and maintenance personnel. In addition, management personnel will need more technical knowledge for appropriate management decision making. Therefore, short-term training programs at local levels have been organized.

Finally, regional mechanization centers have been established. Because of the diversity in local conditions, China's regions have different mechanization needs and problems. Thus, planning and scientific experiments for mechanization have to be decentralized to some extent so that regional variations can be taken into account. On the other hand, within a region, centralization of research and development will increase efficiency and coordination by permitting a greater concentration of resources and by avoiding duplication of efforts. For these reasons, 16 new mechanization centers have been established in different regions to coordinate research and development in mechanization.[16]

In 1979 the Chinese leaders readjusted their development priorities and scaled down their ambitious modernization programs because of growing economic imbalances and financial difficulties. As part of the readjustment, the objective of "basic mechanization of agriculture by 1980" was abandoned.[17] Since then, the Chinese leaders have become more aware of the economic obstacles to mechanization that we shall discuss below and have not stressed mechanization as the major means to agricultural modernization. However, they still regard mechanization as the long-term solution to China's low productivity in agriculture.[18]

Evaluation

The basic flaw in the new measures adopted since 1977 to promote mechanization is that they are concerned only with the supply side of mechanization—that is, how to produce more and better machines and operate them efficiently. An equally important question that has not been adequately addressed by the Chinese leaders has to do with the demand side of the issue: Do the collectives want to mechanize and can they afford those expensive machines? The answer is not an unequivocal "yes" for the following reasons:

1. In most of China's farming areas, labor is abundant but arable land is very scarce. Thus, the existence of surplus labor in many communities is inevitable. In late 1981 the government predicted that "with the introduction of the system of responsibility in agricultural production, surplus labor in the countryside is expected to be as high as 50 percent."[19] Agricultural mechanization will further increase the surplus labor by releasing some laborers from farm work. Therefore, unless the collectives can find alternative employment for the displaced labor, the collectives will be reluctant to mechanize.[20] Alternative employment opportunities depend on the development of rural industries and sideline production. These, in turn, depend on various factors, such as the availability of local raw materials and proximity to urban markets, which are beyond the control of the collectives. It is impossible for them to send their redundant labor to the cities for temporary employment—the government bans the rural-urban migration because the cities are already suffering from unemployment and overcrowding.

2. Because the major justification for agricultural mechanization in many areas is to facilitate triple cropping, the question naturally arises: Does triple cropping increase total output and income? There is evidence that in some areas triple cropping reduces total output because it gives lower average yield per crop due to the fertilizer constraint. For example, in Songjiang County near Shanghai, triple cropping has eliminated the green manure crop, thereby reducing the supply of organic fertilizers and increasing the cost of chemical fertilizers. In addition, the labor requirement for farming has been greatly increased, adversely affecting sideline production. As a result, total output and net income have declined. For these reasons, the county has gone from triple cropping back to double cropping.[21] Similarly, triple cropping has reduced farming income in Wuxi county, Jiangsu Province, a model area in agricultural mechanization. The peasants there would like to return to double cropping, but have been prohibited from doing so by the local party leadership.[22] These examples throw doubt on the wisdom of mutiple cropping in some areas as well as the need to mechanize to relieve the labor shortage that multiple cropping creates.

3. Even if the collectives find it desirable to mechanize farm work, the high cost of machinery presents another problem. Given the low level of rural income and the fact that the collectives are responsible for the cost of mechanization, farm machines are often prohibitively expensive for the collectives to purchase. Collectives often have to use up all their savings to pay for the machines.[23] Although the government gives financial assistance to poorer collectives, the amount of aid has been very small relative to the cost of mechanization.[24] The problem is aggravated by the fact that most farm machines are specific in their functions and remain idle for most of the year unless various attachments are purchased. It is partially for this reason that tractors are often used for transportation purposes.[25] Futher, the cost of machine maintenance and repair is very high because the machines are not yet standardized and the parts are often unavailable. Thus, unless the government increases substantially its subsidy to the collectives for mechanization, a high level of mechanization will remain financially beyond the reach of many collectives for a long time to come.

These problems help explain why the actual pace of mechanization has been far below the original official target. In 1978, only about 33% of the farmland was plowed by tractors.[26] In early 1980, this figure rose only to 40 percent.[27] In 1981, it was 42 percent. In the same year, transplanting machines were used on only 0.7 percent of the paddies, and harvesters on only 2.6 percent of the grain land. Machine sowing was practiced on 13 percent of farmland that was not in paddies.[28] Many observers have attributed the increases in tractor plowing in 1980–81 to the fact that the incomes of many rural households have significantly increased because of the implementation of more liberal agricultural policies and the widespread adoption of the agricultural responsibility system.[29] This is another concrete evidence of the importance of the demand factor in China's agricultural mechanization.

Naturally, there are exceptions to the above problems. Surplus labor does not exist in sparsely populated regions, such as much of the northeast, where many of the state farms are located. Nor is the cost of mechanization burdensome to prosperous collectives or to the state farms where the government finances the cost of mechanization. Consequently the state farms have attained a high level of mechanization and labor productivity and have been publicized as models of modern agriculture. It is inappropriate and misleading, however, to suggest or imply that the mechanization experience of the state farms can be applied to other farming areas.

Interestingly, the American Rural Industry Delegation that visited China in 1975 predicted in their trip report that "China will have a *highly* mechanized agriculture in the not-too-distant future" (italics mine). The basis of their prediction is the "agricultural and related industrial achievements that we saw during our trip,"[30] that is, the aggregate supply capability of the economy. Like the original official Chinese target for agricultural mechanization by 1980, this prediction has not and will not come true because it ignored the demand side of the issue.

Rural Energy

Agricultural machines require power and fuel to operate. Thus, a high level of agricultural mechanization will be impossible without rapidly increasing the supply of energy to agriculture. Beside farm operations, many other rural activities, such as construction, rural communication and transportation, rural industries, and sideline production, will also require more energy in the process of agricultural modernization.

However, agriculture will have to compete for the limited energy supply with other sectors of the economy—industry, transportation, construction, and defense—all of which require more energy for their modernzation goals. In the past, in spite of the fact that the rural population constitutes at least 80 percent of the population, China's rural areas have received a relatively small share of the nation's energy supply. In 1980, for example, China's countryside consumed only 40 percent of the total energy and 12 percent of the total electricity of the nation.[31] Thus, the Chinese planners will have to increase not only the total supply of energy but agriculture's share of the supply as well, if China is to achieve agricultural modernization.

Traditional Fuels

For the vast majority of China's 170 million peasant households, the traditional fuels—firewood, crop byproducts (straw, stalks, vines), grass, and dung—are still the most important sources of household fuels. However, even these have been in short supply; every year, for two months or more, more than 70 million rural households have a shortage in the supply of the traditional fuels.[32] There are three main reasons for this fuel shortage. (1) Historically, the growing demand for food due to population growth has resulted in the expansion of farmland and a low and declining percentage of forest cover in China. (2) About 40 percent of the country's total supply of crop byproducts is used as animal feed, industrial raw materials, construction materials, and fertilizer.[33] (3) The direct burning of crop byproducts is a very inefficient way of utilizing their stored solar energy because only 10 percent of the energy is exploited.

As a result of the fuel shortage, the government has had to allocate some coal for rural household use, thus reducing the supply of coal to industry. Many peasants are forced to fell trees illegally and cut or uproot grass for fuel. This has led to increased deforestation and soil erosion in large areas. To alleviate the shortage, the Chinese belatedly started a large reforestation program in the late 1970s.[34] To increase the energy exploitation rate of crop residues, the government is encouraging the construction of biogas digesters to make methane out of these materials. This will be discussed later in more detail.

Electricity

The most important source of energy for agricultural production is electric power. Electricity is primarily used for irrigation, grain processing, preparation of fertilizers and animal feed, and lighting. China's rural electrification did not begin until 1958, as a byproduct of the construction of irrigation projects of the Great Leap Forward. Hydroelectricity was to be the major power source for small rural industries. Although there was a great increase in the number of small hydroelectric stations in 1958 and 1959, the total generating capacity installed by the end of 1959 was only 400,000 kilowatts, about 10 percent of the planned target (see Table 5.2).[35] Thus the rural electrification plan of the late 1950s failed, along with the failure of the massive water conservation projects.

After the Great Leap Forward, the emphasis of rural electrification was shifted from building small rural hydroelectric stations to building transmission lines to transmit electricity from the cities to the rural areas. One reason for this was that the severe droughts in many parts of the country reduced the efficiency of the small hydroelectric stations. Also, there was underutilization of electricity in the urban areas because of the industrial recession in the early 1960s.

The new government policy was to give priority to national economic recovery.[36] Thus in 1964, the total generating capacity of the rural power stations remained at 400,000 to 500,000 kilowatts, which supplied only 40 percent of the total electricity used by the rural areas. The other 60 percent came from the national power networks.[37]

During the Cultural Revolution decade, both thermal and hydroelectric stations were stressed. In the upper and middle reaches of the Huanghe, in the Changjiang Basin, and along the tributaries of the Changjiang, more than 30 large- and medium-size hydroelectric power stations were constructed by 1975, adding more than 4 million kilowatts generating capacity to the nation's total.[38]

TABLE 5.2
Small Rural Hydroelectric Stations, 1949-80

Year	Number of Stations	Total Capacity (1000 kilowatts)
1949	57[a]	5
1952	98[a]	8
1954	114[a]	8
1956	240[a]	12
1958	4,878[a]	152
1959	n.a.	400
1960	9,000	250-520
1965	n.a.	400-500
1972	35,000	n.a.
1973	50,000	n.a.
1975	60,000	n.a.
1977	81,000	4,320
1978	87,000	5,270
1979	94,214	6,340
1980	90,000-98,214[b]	6,600-7,100

Sources: Leslie T. C. Kuo, Agriculture in the People's Republic of China (New York: Praeger Publishers, 1976), p. 237 for 1949-58, 1960, 1972-75; Kang Chao, Agricultural Production in Communist China, 1949-1965 (Madison: University of Wisconsin Press, 1970), p. 140 for 1959-60, 1965; Renmin Ribao, February 1, 1979, p. 1 for 1977-78; Renmin Ribao, January 18, 1980, p. 1 for the derivation of the 1979 figures; Beijing Review, No. 7, February 16, 1981, p. 6 & Renmin Ribao, November 28, 1980, p. 3 for 1980.

n.a. = not available.

[a]Includes medium-size rural hydroelectric stations.

[b]Beijing Review, No. 7, February 16, 1981, p. 6 gives, for 1980, an increase of "more than 4,000" small stations over 1979, with a total of 90,000. Other Chinese sources have referred to the number of small stations in 1980 as totaling "more than 90,000." Thus the figures 90,000-98,214 are given here to indicate the range for the actual number and the possibility that the Chinese figures for 1977-78 may not be accurate.

More important to the rural areas was the proliferation of small hydroelectric stations, especially after 1970, built by county governments and the rural collectives. As shown in Table 5.2, their number nearly quadrupled between 1960 and 1972 and reached more than 90,000 by the end of 1979, with a total generating capacity of 6.3 million kilowatts. In 1979, these small stations generated 11.9 billion kilowatt-hours, which constituted about one-third of the total electricity used by the rural areas.[39] In 1980, they provided 40 percent of the electricity used in agriculture.[40]

Although the utilization rate of the small hydroelectric stations, in terms of the number of hours utilized per kilowatt per year, is lower than that of the large- and medium-size stations, the small stations have become indispensable to a large number of rural communities in their agricultural production and modernization. More than 1,400 counties out of a total of some 2,000 counties in China have small hydroelectric stations; 720 of them rely primarily on their small stations to supply electricity.[41] In Sichuan, 68 percent of the counties rely primarily on the small stations for electricity.[42]

China's potential for further expansion of small hydroelectric stations is excellent. The nation's hydro resources suitable for small hydropower stations are estimated at 150 million kilowatts, of which about half can be utilized. As the current utilization rate is only about 10 percent, there is much scope for future expansion.[43]

There are various advantages to using small hydroelectric stations for rural electrification. These stations are built primarily by the rural collectives with local funds, with the state providing technical assistance and equipment; thus the stations are dependent on local initiatives rather than on the availability of state funds. The construction of the stations is labor intensive, does not require much sophisticated equipment, and takes a relatively short time. The benefits of the approach are not limited to electricity generation—the reservoirs are also used for local water conservation, drainage, irrigation, river navigation, and breeding of fish and other aquatic products.[44]

In spite of the progress made in rural electrification, particularly in the construction of small hydroelectric stations, only 87 percent of the communes, 62 percent of the brigades, and some 50 percent of the production teams had electricity in early 1980.[45] The rural areas used only 37 billion kilowatt-hours, or about 12 percent of total electricity generated in 1980, which was far from adequate. If the electricity consumption of county-run, agriculture-related industries is included, the percentage is about 25 percent.[46]

Nationwide, hydroelectricity amounted to only 17 percent of

the total electricity generated in 1979, with the rest coming from thermoelectric plants. This is a decline from 25 percent in the earlier years and reflects a gross underutilization of the hydro-energy resources of the country. It has been estimated by the Chinese that China has a potential hydropower of 680 million kilowatts, of which 370 million kilowatts can be exploited.[47] In 1979, only 2.5 percent of this exploitable capacity was utilized. The scope for future expansion of hydroelectric power is therefore enormous.

The reason hydroelectric power is less important than thermal power in China lies in the Chinese planners' belief that hydroelectric stations require much more initial investment and have a longer gestation period than thermoelectric stations. This is generally true, but thermoelectricity will become more expensive if additional coal mines and/or railroads have to be constructed, as Chinese energy experts believe may be the case in the future. Consequently, in recent years it has been recommended that priority in the future development of electric power be given to hydro rather than thermal plants.[48]

Coal

Coal is currently used to generate the bulk of the electricity in China. In 1979, it accounted for 62 percent of the electricity generated, as compared with 21 percent for oil and 17 percent for hydro plants.[49] Coal production in China has been growing quite rapidly since 1949, rising from 32 million metric tons in 1949 to 310 million tons in 1970 and 635 million tons in 1979 (Table 5.3), making China the third largest coal producer in the world. Coal production is currently concentrated in the northeast, the north, and the northwest where hydro resources are less abundant than in the south. In 1977, one-third of the total coal output was produced by some 20,000 small mines located in about half of China's counties.[50]

China's coal reserves are estimated by the Chinese at 600–650 billion metric tons. Most of these, however, are located in the relatively inaccessible northwest and southwest, with more than 200 billion metric tons in Shanxi province alone.[51] In addition, China's coals are generally of low quality—one kilogram of coal produces about 5,000 kilocalories in China as compared with

TABLE 5.3
Output of Coal and Crude Oil, 1949-80
(million metric tons)

Year	Coal	Oil	Year	Coal	Oil
1949	32	0.12	1965	220	10.96
1950	43	0.20	1966	248	14.07
1951	53	0.31	1967	190	13.90
1952	66	0.44	1968	205	15.20
1953	70	0.62	1969	258	20.38
1954	84	0.79	1970	310	28.21
1955	98	0.97	1971	335	36.70
1956	110	1.16	1972	356	43.07
1957	131	1.46	1973	377	54.80
1958	230	2.26	1974	384	65.77
1959	330	3.70	1975	427	74.26
1960	280	5.10	1976	448	83.61
1961	170	5.19	1977	550	93.64
1962	180	5.75	1978	618	104.00
1963	190	6.36	1979	635	106,15
1964	204	8.65	1980	620	105.95

Sources: Joint Economic Committee, Chinese Economy Post-Mao, p. 367 for 1949-76; Beijing Review, No. 27, July 6, 1979, p. 37; No. 19, May 12, 1980, p. 13 & No. 19, May 11, 1981, p. 24 for 1977-80.

more than 6,000 kilocalories in the industrialized countries; less than 20 percent of China's coals are washed and separated from noncoal substances.[52] Finally, due to low investment in the past, most coal mines have a low level of mechanization and small scale of operation, resulting in inefficient production.

In 1977-78 when the Chinese leaders first launched their modernization programs, their target for the coal industry was to increase its annual output to one billion metric tons by 1987. This was to be accomplished by increasing the mechanization of existing mines and the building of large new mines with the aid of imported technology and equipment.[53] A more balanced distribution of the coal mines was also to be achieved. Since 1979 when the Chinese government readjusted its economic priorities and downscaled its investment projects, that output target for the coal industry has not been mentioned any more. As of late 1981, the short-term objectives are to ensure an annual output level of 600 million metric tons while improving the organization and efficiency of existing mines, to expand and mechanize the existing ones, and to improve the quality and industrial utilization of coals. Shanxi, East China, Northeast China, and Henan will be the key areas of development.[54] Apparently, the Chinese authorities have decided that it is more economical to concentrate on areas that have better mines and better transportation facilities.

The prospects of China's coal industry will depend on the availability of financial resources for new investment and technological improvement. Increased production of coal is important, not only for use in industry and in thermal power plants, but also for releasing more oil for export.

Oil

Oil is another important source of energy for agriculture. It aids agriculture indirectly by producing 21 percent of the total electricity generated. More importantly, diesel fuel is used directly to power farm machines. Diesel fuel for agricultural use is subsidized by the government (see Chapter 6), but the supply is limited. In 1980 China supplied only some 50 kilograms of diesel fuel for each horsepower of agricultural machinery in use, which is sufficient for only 50 days of field operation.[55] This is one of the major reasons for the low rate of utilization, and hence the high cost of operation, of farm machines.

The oil industry has been one of the fastest growing industries in China. During the First Five-Year Plan, crude oil output grew from 436,000 metric tons in 1952 to 1.46 million metric tons in 1957, an average annual growth rate of 27.3 percent. This was accomplished by expanding the productive capacity of existing oil wells and by stepping up the rate of drilling. The investment priority of the government was on exploration. This effort started to pay off in 1959–60 when proved reserves increased from less than 30 million metric tons to more than 100 million metric tons in 1957, and to 200 million metric tons in 1959–60.[56] In particular, the huge reserves of the Song-Liao Basin in Northeast China were discovered, and the principal oilfield there, Daqing, began production in 1960. As a result, crude oil output increased rapidly from 5.1 million metric tons in 1960 to 28.2 million metric tons in 1970 (see Table 5.3). The Daqing discovery was followed by the discoveries of the Shengli and Dagang oilfields in North

China, which were brought into production in 1967 and 1973, respectively. In 1973, China began to export crude oil to Japan. The crude oil output level continued to rise rapidly throughout most of the 1970s. In 1979, however, the annual rate of increase slowed to a mere 2 percent, and China indicated to Japan that it would have difficulties in honoring export commitments. In 1980, crude oil output level did not increase at all. However, since China is still exploring a number of new oil fields, including some offshore, and there are indications that the exploration will be successful,[57] it is likely that the slowdown will be temporary.

Western estimates of China's crude oil reserves vary greatly. One study concludes that the reserves are likely to fall between 3 and 10 billion metric tons. The upper limit of this range would make China the third richest oil region of the world, just behind the Middle East and the Soviet Union.[58]

In the foreseeable future, it is doubtful that the supply of oil to agriculture will be substantially increased. Since the mid-1970s, China has relied on oil export as one of the major sources of foreign exchange earnings to finance its imports of technology. China will certainly continue to export oil in the future, rather than increase its oil supply to agriculture. Also, other sectors of the economy, particularly industry and transportation, will have increasing demand for oil during the process of modernization, further limiting the supply available to agriculture. The Chinese leaders are likely to stress the development of rural energy sources—small hydropower stations, small coal mines, and biogas digesters—which utilize rural resources and minimize the competition for oil between agriculture on the one hand and export, industry, and transportation on the other hand.

Biogas

A relatively new source of energy is biogas (or "marsh gas" in Chinese terms), which contains about 70 percent methane and 30 percent carbon dioxide. It is generated in sealed underground digesters through the fermentation of organic matter, such as animal and human wastes, crop residues, grass, and leaves (see Figure 5.4).

Initially introduced in the 1950s, biogas was popularized in Sichuan in the early 1970s for home cooking and lighting. By

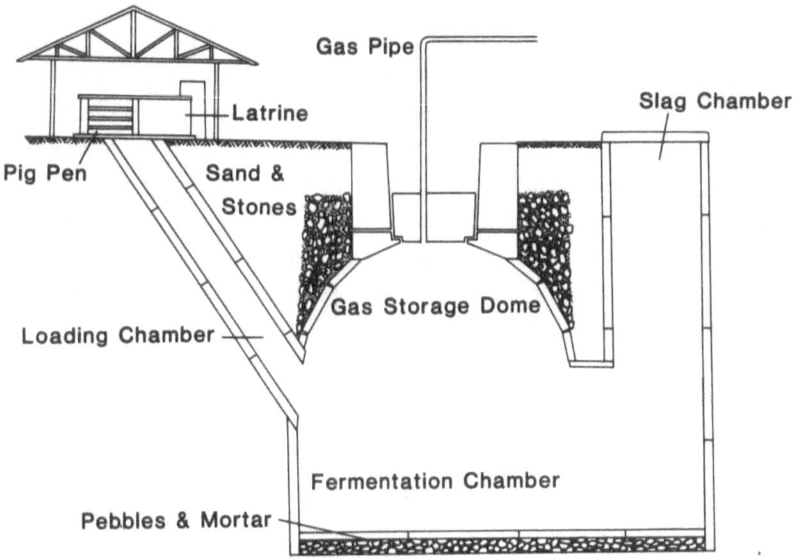

Figure 5.4. Biogas digester. Figure shows a design in wide use in Sichuan. A digester with 10 cubic meter capacity (2.5 meter in dome diameter) generates enough biogas for household use. (Based on *Nongchun Ban Zhaoqi*, Beijing: Science Press, 1975, p. 24)

1974, it was also used for the generation of electricity. In 1980, there were more than 7 million digesters with 8–10 cubic meter capacities each throughout China's countryside, used by 30 million peasants.[59] In addition, there were some 36,000 larger digesters built by communes, brigades, state farms, wineries, bakeries, and food-processing factories for generators, water pumps, irrigation sprinklers, rice threshers and dryers, and mills.[60] All of these make China the world leader in the development and exploitation of biogas.[61]

The development of biogas is uneven in China. Sichuan has 63 percent of the digesters, followed by Jiangsu (9.4 percent), Zhejiang (5.4 percent), and Shandong (5 percent).[62] Since the fermentation efficiencies are higher in the warmer southern areas, it can be expected that the southern provinces will continue to lead the nation in the development of biogas.

The benefits of biogas production are numerous. Economically, it is an efficient way of exploiting the energy of organic waste matters. In terms of construction cost, a 10-cubic-meter digester requires 40–50 yuans or less of building materials (bricks, stone, and cement), which are economically within the reach of a peasant family.[63] The biogas produced from such a digester is sufficient for the fuel needs of a peasant family, and thus the building cost will be recouped within a year. Biogas can also be combined with diesel fuel for powering diesel engines, saving about 70 percent of the diesel fuel, according to the Chinese experience.[64] Also, electric power stations using biogas are easy to construct and maintain. According to Chinese estimates, the initial investment for generating one kilowatt is only one-half to one-third that of the small hydropower stations.[65] One final economic benefit is that the sludge removed from the disgesters makes a rich nitrogen fertilizer.

From the point of view of health and environment, the burning of biogas is cleaner and more convenient than that of solid fuels. The fermentation of animal and human wastes also helps control diseases.

All of these considerations suggest that the prospects of continuous growth in biogas use in rural China are very good and that biogas will go a long way in supplementing hydropower as an important source of energy for rural use.

Other Sources of Energy

Another promising source of energy is solar energy. Two-thirds of the country has more than 2,000 hours of sunshine. At present, however, the use of solar energy is still at an experimental stage, confined to the use of solar stoves and solar water heaters for cooking, food drying, and heating of animal pens. In 1980, China had a total solar collector surface of only 60,000 square meters.[66] Other forms of renewable energy, such as wind power and geothermal power, are also only at the experimental stage.

Plant Protection

Pests that damage crops include birds, insects, mammals, and other animals; plant diseases; and weeds. One estimate puts annual worldwide crop losses from insect damage at 13.8 percent, from diseases at 11.6 percent, and from weeds at 9.5 percent.[67]

Thus, the protection of plants from pests can be significant in increasing food supply.

China's most common insect pests and plant diseases include locusts, rice borers, armyworms, corn borers, cotton aphids, red spiders, pink bollworms, wheat-juice-sucking insect larvae, wheat smut, wheat nematodes, and black rot on sweet potatoes. In total, there are more than 30 major nationwide plant diseases and pest insects and more than 40 regional ones. To protect plants from these pests, China has adopted multifaceted pest-control programs, which rely heavily on the rural organizations.

Available evidence indicates that China's plant protection programs have been quite successful. For example, locusts have destroyed crops and devastated large areas throughout China's history. Between 700 B.C. and 1949, over 800 locust plagues were recorded. By the end of 1962, the area affected by locusts had been curtailed by more than 60 percent.[68] By 1978, the Chinese indicated that they had brought the locust problem "completely under control."[69] This was accomplished by the use of insecticides and the transformation of locust breeding grounds. Yellow wheat rust has been virtually eliminated from China's main wheat-producing areas through the use of better seed strains that resist the disease, along with the use of insecticides and better cultivation methods. Damage from rice borers has been reduced from 10 percent of the harvest in the past to 1 percent or less.[70] The armyworm, traditionally regarded by the peasants as the "divine insect," has also been brought under effective control.[71]

There are two major aspects to China's plant protection programs, the organization and the technology. Both are important to China's success in plant protection.

Organization of Pest Control

The organization of pest control encompasses the national, provincial, and county institutions and units of the rural collectives. At the national and provincial levels, research institutions conduct research and train pest-control personnel from the lower levels. The Plant Protection Institute of the Chinese Academy of Agricultural Science is the national center for forecasting the outbreak of pests that migrate from province to province, such as the armyworm.[72] The provincial institutes also collect field data for experimentation and forecasting. The center for most pest-control activities in the countryside, however, is the county disease and insect forecasting station or the county plant protection unit with its team of trained plant protection workers. These workers maintain observation plots at the county experimental farms and monitor the development of insects and diseases. In case of imminent outbreaks of insects or diseases, they alert all the communes in the county and recommend appropriate actions.[73]

Below the county, the communes have insect and disease forecasting stations, and the brigades and teams have plant protectors in the field. Together with the county plant protection workers, they form a countywide network for plant protection. Field information and technical experience are exchanged in the network. In case of pest outbreak, the brigade and team protection workers report daily to the commune stations, which, in turn, report to the county center. The county evaluates the information and suggests control methods for the county. The communes and brigades also make their own forecasts, as each locality has different natural conditions and weather.[74]

Pests infest a geographic area rather than individual farms, so the collective nature of China's agriculture makes it easier for farmers to carry out and coordinate pest-control activities without being dependent on government actions. In an alternative setting of small-scale private farming, such as in pre-1949 China or in most of the contemporary developing countries, individual peasants would not have the incentive to undertake independent pest-control activities because such activities would not be effective. Thus, government pest control is the only solution.

Pest-Control Technology

Technologically, the Chinese have progressed in developing and adapting different methods of pest control which, when used in combination, are not only effective in reducing pest damage to crops but are also environmentally sound, safe for people, and low in cost. This is the approach of integrated pest control. The major pest-control methods being used are insecticides, insect trapping, biological control, host resistance, and cultural and environmental control.

Insecticides. The use of insecticides is an important method of pest control. The Chinese are aware of the associated problems of insect resistance, the rise of secondary pests, possible soil and water pollution, and injury to human health, and have become more selective in insecticide use in order to maximize its actions on the pests but minimize its impact on nontarget insect species and on humans and animals.

Organophosphorus insecticides are the basic insecticides used in China today. Of these, trichlorfon is in widest use because of its safety, lack of persistence, environmental degradability, and low cost. It is used on cotton, tobacco, rice, cabbage, and other vegetable crops. The other insecticides, in descending order of importance, are dichlorvos, dimethoate, phosmet, fenitrothion, phosphamidon, and malathion. Dichlorvos is used on cotton, soybeans, and peaches. Dimethoate is the basic insecticide used on rice.[75]

The organochlorine insecticides, formerly in wide use in China, are being phased out because of their environmental persistence, the development of insect resistance to them, and hazards to human health. Thus the use of DDT on fruits and vegetables is prohibited, and it is no longer used on rice because the rice paddy borer is resistant to it. The use of benzene hexachloride (BHC), formerly the most popular insecticide in China, is also prohibited on fruits and vegetables.[76]

Biological Control. Biological control utilizes the natural enemies of pests—parasites, microbial agents (bacteria, fungi, viruses), and predators—to reduce the pest population. The systematic use of biological control to conquer plant pests and diseases in China began in the mid-1950s, but remained on a small scale until 1972. Since then, it has been rapidly spreading because it is effective and inexpensive, does not pollute the environment, and is harmless to the peasants and farm animals.

The most popular method of biological control is the mass release of *Trichogramma* (tiny parasitic wasps). They are released in farms and forests to control the European corn borer, pine caterpillar, rice leafroller, sugarcane borer, and other Lepidoptera in cotton in various parts of the country.[77] In 1977, the parasite was used on 660,000 hectares. Its efficiency rate is reported to be 70 to 90 percent.[78] To produce large quantities of *Trichogramma* for release, entomologists breed them on the eggs of giant silkworm moths and rice grain moths.

The release of *Trichogramma* can reduce rice leaf damage from 20 to 1 percent and cut the use of insecticides by half.[79] In use against cotton borer in Shanxi, the cost of pest control per mou was reduced from 5.36 yuan before 1978 to 0.37 yuan in 1979.[80] The cost of *Trichogramma* production is lower than that of insecticides because only local raw materials and labor are used. As a side benefit, more silk cocoons are produced.

Beauveria bassiana, a bacterium isolated from an endemic infection of the European corn borer, is used to control corn borer (on corn, sorghum, and millet) and a pine moth. The efficiency rate against the corn borer is about 80–90 percent. The culture is produced at the commune level by using local raw materials such as wheat bran, rice powder, cornstalk powder, and compost humus.[81] Thus, it has the advantages of efficiency, low cost, and simplicity of production.

In addition to the above, *Bacillus thuringiensis* (B.T.), a bacterium, is used on cabbage to control diamondback moths. The larvae of a hymenopteran parasite are used to control the overwintering larvae of the pink bollworm in cotton storehouses.[82] Predators such as ladybeetles and lacewings are used against cotton pests; more recently, spiders are being used to control rice pests.[83]

Ducks are often used as an alternative to insecticides in rice paddies to control rice insects and weeds. This is an efficient method because a 250-gram baby duck can eat more than 400 insects in an hour, according to field observations.[84] Further, this has the salutary effect of increasing the commune's poultry production and income. As the herding of ducks from field to field is labor-intensive, it provides additional employment opportunities to commune members.

Host Resistance. Host resistance consists of breeding crop varieties that are resistant to diseases and insect pests. For example, varieties of wheat resistant to stalk rust have been popularized in the northeast, and breeding programs for resistance to insect pests are being conducted on cotton and wheat in Shaanxi province.[85] In general, however, the American Insect Control

Delegation to China in 1975 found breeding for resistance to insect pests to be in its infancy. There is little work on plant resistance to the European corn borer, soybean pod borer, aphids, and to insects of small grains and sorghum. Instead, more emphasis is placed on breeding for resistance to all major plant diseases occurring on these crops.[86] The latter is undertaken in conjunction with regular plant breeding for developing improved varieties and is carried out both at research institutes and within the communes.

Other Control Methods. Other methods of control include cultural control, environmental control, and insect trapping. All of these are essential parts of China's integrated control. Cultural control involves modifications in agronomic practices to control certain pests and is based on knowledge of the pest's biology and ecology. According to some Chinese scientists, cultural control constitutes the base of their integrated control program.[87] Examples include winter plowing and winter flooding of rice fields to kill rice borers; rational crop rotation to control cotton wilt, and proper control of water and fertilizer supply to alleviate rice blast.[88] In addition, summer weeding in rice paddies and winter pruning of citrus trees increase plant vigor and reduce pest problems.[89] Finally, there is strict seed certification by county officials for shipment of seeds out of the county. This limits the spread of seed-borne pathogens. The American Insect Control Delegation was favorably impressed with China's success in cultural control.[90]

Environmental control involves broader transformation of the pests' environment to destroy their hibernation and breeding sites. For example, China has had impressive success in controlling locusts, accomplished by the transformation of locust breeding grounds in North China through water management, reforestation, and land reclamation.[91]

Insect trapping also plays an important role in plant protection. Light traps, especially the fluorescent blacklight traps, are popularly used for insect control as well as for forecasting of insect abundance.[92] Bail pails are also used in some areas.[93]

Summary

Plant protection in China is multifaceted and integrated. It is implemented nationwide through the extensive commune system. The methods used are generally labor-intensive, thus effectively mobilizing the abundant labor of the country. Overall, China's plant protection is highly successful. In the words of the American Insect Control Delegation to China, "interest in and support for plant protection and particularly in development of integrated control was very impressive in China. It has a much higher priority than in the United States and perhaps more than in any other part of the world."[94]

6
Financing Agricultural Development

Agricultural development requires not only government support for research and development of better technology, but also resources at the farm level to purchase modern inputs such as chemical fertilizers and better seeds and to finance fixed capital projects. Because China's agriculture belongs to the collective sector, the collectives are responsible to a large extent for their own investment. The sources from which they finance their own investment include the public accumulation funds, collective savings and investment in kind, and profits from collective enterprises.

The government also affects agricultural investment in various ways through its financial policies. Directly, the government invests capital in agriculture and provides financial assistance to some communes and brigades. The banking system extends agricultural loans to the collectives for production purposes. Indirectly, the government affects the size of the collectives' incomes and investment funds through its price and tax policies.

In short, government financial policies toward agriculture determine the level of state resources allocated by the government to agriculture. Because similar financial instruments also exist to allocate state resources to other sectors of the economy, an examination of government financial policies is a good way to study the development strategy of the country and the relative priorities that the government attaches to various sectors of the economy.

Portions of this chapter have been published in Robert C. Hsu, "Agricultural Financial Policies in China, 1949–80," *Asian Survey*, Vol. 22, No. 7 (July 1982).

Government Price Policies

Government price policies are probably the most important and powerful policy instruments that the government can use to affect the incomes of the agricultural sector. The prices of most agricultural products in China are controlled by the government. This control is considered to be essential to the operation of China's planned economy because agricultural prices affect the cost of living and thus industrial wages; they also affect the cost of industrial raw materials and thus the prices of some manufactured products.

For purposes of pricing and market distribution, agricultural products are officially classified into three categories according to their importance to the population. Although the items in each category have changed slightly since the 1950s, the principle of classification and the basic items in each category have remained the same.

1. The "first category products" comprise those that are essential to the population; they include the major crops, such as grains, cotton, soybean, rapeseed, and peanuts. Since the early 1950s, under the system of "planned purchase," annual compulsory delivery quotas have been set for these products for each province by the State Planning Commission. The size of the quota is determined by the needs of the consumers as well as the capabilities and needs of the producing areas. The provincial quotas are ultimately disaggregated into quotas for the production teams through the local planning process. Purchase prices for these products are set by the government. The products are then rationed to the consumers at retail prices, which are also controlled by the government.[1]

For purchases over and above the quotas, higher purchase prices are paid. The government also sets targets for the production teams for the "above-quota purchase." However, there are no penalties for failure to meet the targets.[2] If the teams or individual peasants wish to sell more grain than the target for the above-quota purchase to the state, the price will be negotiated between the sellers and the local food bureaus according to supply and demand conditions.[3] In theory, the state grain departments may not refuse offers to sell them more grain.[4] In practice, because of warehouse shortage or other problems, some local grain departments have refused to purchase any grain over the above-quota target or have offered very low prices, affecting adversely the peasants' incentives to produce.[5]

2. The "second category products" are those that are important to the national economy for consumption or export; they include major medicinal herbs, ramie, hemp, sugar cane, tea, wool, cowhide, live pigs, and tung oil. They are purchased under the system of "unified purchase" by the state commercial departments at prices determined by the government. Quotas are also set by the Ministry of Commerce for the collectives to sell these products to the state. However, the collectives can sell the surplus over the quota on the open market at market prices. Also, the products are not rationed to the consumers.

3. The "third category products" encompass those outside the first two categories; these are primarily minor local products and sideline products, such as minor medicinal herbs, poultry, and animal byproducts. They are outside the quota system and may be sold on the rural free markets or exchanged through negotiations. However, because rural free markets were closed during some periods of ideological excesses, as discussed in Chapter 3, and because these products had little local demand in some areas, the peasants often had to sell them to state-run supply and marketing cooperatives at "negotiated" or posted prices.

In addition to agricultural prices, the prices of industrial products used in agricultural production, such as chemical fertilizers, kerosene, diesel oil, and farm machines, also affect the peasants' cost of production and income. Industrial prices are determined by the state, generally on the basis of the cost-plus principle, that is, the average cost of production of the industry plus a certain percentage of planned profit.[6] When these products are sold to the collectives, the state may include some subsidy to lower the prices.

In the 1950s, when China adopted the Soviet model of development, which emphasized industrial development at the expense of agriculture, agricultural prices were set at a very low level but industrial prices were set at a relatively high level. Thus the peasants were exploited by the double squeeze of the "price scissors."

TABLE 6.1
Purchase Prices for Farm and Sideline Products, 1957-75
Hanshou County, Hunan Province
(yuan per kilogram)

Year	Wheat	Ginned Cotton	Jute	Pork	Fish	Egg
1957	0.146	1.57	0.74	0.764	0.44	0.80
1965	0.228	1.77	0.84	0.92	0.58	1.24
1975	0.26	2.10	0.84	0.92	0.78	1.32

Source: Peking Review, Nos. 32-33, August 9, 1976, p. 26.

TABLE 6.2
Prices of Industrial Goods Used in Agriculture, 1957-74
Hanshou County, Hunan Province
(yuan per kilogram)

Year	Chemical Fertilizer	Pesticide	Diesel Oil
1957	0.42	1.35	0.264
1965	0.38	1.15	0.164
1974	0.29	1.04	0.164

Source: Peking Review, Nos. 32-33, August 9, 1976, p. 26.

This naturally reduced the peasants' incentives to produce as well as their ability to invest. Since the late 1950s, as China abandoned the Soviet model and developed its own strategy of development, prices began to be changed in favor of agriculture, as can be seen from Tables 6.1, 6.2 and 6.3. These changes were further justified ideologically in terms of strengthening the "worker-peasant alliance," meaning that it is important to reduce the income disparity between the industrial workers and the peasants in order to promote socialist development.

For some products that are sold to both agriculture and the other sectors, preferential prices are given to agriculture. This is the case with electricity and diesel fuel, as shown in Table 6.4. All

TABLE 6.3
Indices of Prices Received and Paid by Farmers, 1950-79
(1950 = 100)

Year	Purchase Prices of Farm and Sideline Products	Prices of Industrial Products Sold in Rural Areas	Terms of Trade[a]
1952	121.6	109.7	110.8
1957	146.2	112.1	130.4
1965	187.9	118.4	158.7
1974	204.5	110.0	185.9
1975	208.7	109.6	190.4
1976	209.7	n.a.	n.a.
1977	209.2	n.a.	n.a.
1978	217.4	109.8	198.0
1979	265.5	109.9	241.6

Sources: Annual Economic Report of China, 1981, part VI, p. 23 for all years except 1974 and 1976-78; FAO Monthly Bulletin of Statistics, April 1981, p. 24 for purchase prices for 1974 and 1976-78; Xiao Zhuoji, "The Law of Price Movement in China," Social Sciences in China, Vol. 1, No. 4 (December 1980), p. 46 for industrial prices for 1978; derived from Nicholas R. Lardy, "Economic Planning and Income Distribution in China," Current Scene, Vol. 14, No. 11 (November 1976), p. 6 for 1974 industrial prices.

[a]Refer to the ratios between purchase prices of farm and sideline products and prices of industrial products sold in rural areas.

TABLE 6.4
Prices of Fuel and Electricity, 1975

Item	Purpose	Price (yuan)
Fuel		
Diesel, kilogram	General	0.40
Diesel, kilogram	Agricultural	0.27
Gasoline, liter	General	1.42
Electricity, kilowatt hour	Household	0.07
	Industrial	0.06
	Agricultural	0.03

Source: American Rural Industry Delegation, Rural Small-Scale Industry in the People's Republic of China (Berkeley: University of California Press, 1977), p. 123.

these price changes and price differentials in favor of agriculture have resulted in improving the terms of trade, or price ratio, between agricultural products and industrial products that the peasants purchase, as shown in Table 6.3.

These changes in prices and the terms of trade in favor of agriculture have often been cited by the Chinese government and some Western scholars as evidence of China's pro-agriculture strategy of development and its industry-support-agriculture policy. There is no doubt that these changes have brought about some redistribution of income from the rest of the economy to agriculture. It is my judgment, however, that the extent of deliberate support given to agriculture through government price changes has been much smaller than the price figures would indicate, for a number of reasons.

First, industrial prices are generally determined on the basis of the cost-plus principle; therefore, industrial prices in the 1950s should be relatively high because of higher costs of production due to smaller scale of production, inferior technology, and lack of experience. By the early 1970s, one would expect industrial prices to decline due to improved technology, skills, and economies of scale. This would lead to some improvement in the terms of trade for agriculture.

Second, where there is clearly an element of subsidy to agriculture, as in the case of electricity and diesel fuel, the extent of subsidy is not as high as the figures would suggest. The reason is that the supply of the subsidized products is very limited in quantity and strictly rationed to agricultural users. This has caused production interruptions because the peasants are not receiving as much of the subsidized products as they need.[7]

Third, as has already been mentioned, some local grain departments have refused to purchase surplus grains at prevailing prices or at reasonable prices. Similar problems have been encountered by the peasants in selling some second and third category products. The reasons varied from lack of storage facilities, to transportation bottleneck, to low profit prospects.[8] Most ironic of all is the case of hogs. It has long been the government's policy to encourage the raising of hogs by the peasants because per capita meat consumption in China has been very low and because hogs produce good organic fertilizer. Yet it has been frequently reported that some local commercial departments refused to buy peasants' hogs at government purchase prices.[9]

These problems stem from the facts that the state commercial organizations have a virtual monopoly in the purchase and distribution of agricultural and sideline products and that the peasants do not have the facilities or connections to sell the products themselves outside of local areas. Thus, when the commercial organizations refuse to purchase their surplus products, the peasants may have no choice but to destroy them.[10]

Finally, in a country as large and as diverse as China, violations of government regulations by lower level bureaucrats for personal gains are not unusual. Thus an "implementation gap" has existed between government policies and their implementations. In state commercial organizations, for example, this has taken the form of price violations by local state stores and rural commercial units, the latest cases occurring in 1979–81. Price violations were so widespread and flagrant that the State Council in November 1979 and December 1980 ordered all commercial departments in the country to carry out general price inspections. Citizen groups ("price sentries") were recruited in 1980 in some large cities to check store prices. Innumerable cases of price violations were uncovered in these inspections. The problem remained unabated throughout 1981. In Beijing alone, 60,225 cases of price violations and tax evasions were confirmed by the authorities.[11] In the cities, the violations took the forms of either unauthorized raising of consumer goods prices, short measures, or disguised price increases by illegally upgrading agricultural products.[12] In the rural areas, it was found that some local purchasing stations illegally reduced purchase prices or arbitrarily downgraded the quality of the products delivered, particularly where harvests were good.[13] There were also cases of arbitrary increases of delivery quotas to avoid the payment of higher prices for the above-quota purchases of good harvests.[14] Clearly, China's "commercial comrades" are no more scrupulous than their counterparts in the capitalist countries and have caused the actual prices to deviate from the official prices, thus cheating both peasants and urban residents.

In short, official price figures exaggerate the improvement in agricultural prices and the extent of government support for agriculture. Hu Qiaomu, China's eminent theoretician, warned in 1978 that the disparity between agricultural and industrial prices was still too large, and that this had become a burden to the peasants because industrial goods were increasingly used in agricultural production.[15] Similarly, some localities still complained in 1979 that "procurement prices for some products do not cover production costs."[16] In 1979, as part of the new government's efforts to increase agricultural production, large increases in state purchase prices of agricultural products were made. The purchase price of quota grain was raised by 20 percent while the purchase price for above-quota grain was increased by 50 percent. On the average, the purchase prices of 18 major agricultural and sideline products were increased by 24.8 percent.[17]

The latest price increases, substantial as they were, might not be sufficient to give the agricultural producers the same rate of return as that earned by state-run enterprises. Certainly the price increases have not been sufficient to narrow the income disparity between the peasants and the industrial workers; instead, the disparity has increased. According to Chinese estimates, per capita income in the rural areas was 117 yuan in 1977 and 170 yuan in 1980; the per capita income of workers in state-owned enterprises was 602 yuan in 1977 and 781 yuan in 1980.[18] It is not surprising, therefore, that after three decades of price changes in favor of agriculture, the Chinese in 1981 still conclude that "the present structure of price in China is in many ways irrational, mainly in the excessive disparity between those of agricultural and industrial products."[19]

Government Fiscal Policies

Government fiscal policies toward agriculture consist of taxation of the rural areas and government expenditures to assist agriculture.

Taxes

The most important tax in the rural areas is the agricultural tax, which is sometimes called the "public grain" because it is col-

TABLE 6.5
Agricultural Tax as Percentages of National Agricultural Output
1949-80

Year	Percentage	Year	Percentage
1949	13.89[a]	1961	6.7[b]
1952	13.2	1970	6
1953	11.94	1972	6
1954	12.43	1974	6
1955	11.65	1975	5
1956	10.76	1980	4
1957	11.51		

Sources: Nai-Ruenn Chen, Chinese Economic Statistics (Chicago: Aldine Publishing Co., 1967), p. 469 for 1949; Joint Economic Committee, China: A Reassessment of the Economy (Washington, D.C.: U.S. Government Printing Office, 1975), p. 363 for 1953-57, 1970-72; Peking Review, No. 26, June 28, 1974, p. 19 for 1961; No. 17, April 26, 1974, p. 5 for 1974; No. 37, September 12, 1975, p. 25 for 1975; Beijing Review, No. 29, July 21, 1980, p. 5 for 1980.

[a]For Gansu province only.

[b]For Tibet only.

lected in grain. It is levied on the output of basic grains, oil-bearing seeds, cotton, and tobacco.[20] The amount of tax is assessed on the basis of "normal" rather than actual yield of grain on comparable land. Once fixed, this norm remains unchanged for a long period of time. Since actual yields have gone up, the agricultural tax as a percentage of actual farm output has declined. As shown in Table 6.5, it has declined from 13.2 percent in 1952 to 5 percent in 1975 and 4 percent in 1980.

Because the agricultural tax rate is not progressive and is based on normal yield, the richer areas, where yields have increased most, pay a smaller percentage of their output as taxes. However, there are provisions in the system to provide relief to the poor collectives. In case of crop failures, part or all of the tax can be forgiven. For poor collectives in economically backward areas, especially in mountainous areas and border regions populated by the minorities, a minimum income level has been established below which no tax will be assessed.[21]

The industrial and commercial enterprises of the rural collectives pay the regular industrial and commercial business tax and a 20 percent industrial and commercial income tax on their annual net profits above 3,000 yuan. Before 1979, the same rate applied to annual net profit over 600 yuan.[22] Since 1979, newly established collective enterprises that have financial difficulties may be exempt from the taxes for two to five years.[23]

The new personal income tax, introduced in 1980, applies only to monthly income in excess of 800 yuan. It was introduced primarily to tax the high salaries and wages of resident foreign employees of foreign corporations doing business in China. Few Chinese would qualify, so the new tax will have no impact on the incomes of the rural areas.

Government Expenditures on Agriculture

Government expenditures on agriculture, as distinct from bank loans to, or purchases from, agriculture, consist of two major categories: grants to communes and brigades for agricultural development, and state investment in agricultural capital construction. More specifically, the government has allocated funds for the following purposes:

1. Aid to communes and brigades for "basic farmland construction." This includes the construction of reservoirs and irrigation canals and the drilling of tube wells for water control and irrigation, thereby expanding the irrigated area with high and stable yields. Nearly 60 billion yuan have been spent on this since 1949.[24]
2. Assistance to poor communes and brigades for the purchase of farm machines and the development of their enterprises and small hydroelectric plants. About 12 billion yuan have been spent for this purpose between 1959 and 1978.[25]
3. State investment in the development of regions for commodity grain, forestry, fishery, and animal husbandry. During the second half of the 1960s, the emphasis of the state investment program was on the traditional grain deficit areas of Hepei, Henan, Shandong, and Jiangsu.

During 1978–79, the emphasis was shifted to Heilongjiang and Jilin for the development of grain-producing farmland and to Inner Mongolia for the development of grazing land.

4. Financing the operation and development of state farms and state ranches. In 1980, there were some 2,900 such units, employing 5 million workers and occupying 4 percent of the nation's arable land.[26] Since 1949, 16 billion yuan have been spent on these state farms and ranches. Because of low yield and poor management, most of them have incurred losses throughout the years.[27]

The percentage of total state expenditures allocated to agriculture throughout the years has been relatively small, as shown in Table 6.6. The percentage was very low, no more than 5 percent, in the early 1950s, but increased somewhat in the late 1950s, increased to 16–18 percent in the early 1960s, then declined and stabilized at little more than 10 percent in the 1970s.

Thus with the possible exception of the early 1960s, agriculture has not received sizable fiscal support. Although it has often been asserted that agriculture is the foundation of the economy, the priority of the state expenditures has been on industry rather than agriculture. This remains true even in the late 1970s and 1980 after the sequence of development priority has been officially reversed from heavy industry first, then light industry, then agriculture to agriculture first, then light industry, and heavy industry last.

In addition to the small size of state expenditures on agriculture, the allocation of funds within agriculture can be criticized. First, from the incomplete figures given above, state farms and ranches have received about 18 percent of the state agricultural expenditures. As these state farms and ranches occupy only 4 percent of the arable land, the state expenditures on them per unit of land are more than four times what has been given to the communes and brigades. From the point of view of raising total food production, it would have been more efficient to allocate more funds to the collectives instead, especially because the state farms have had a long history of low yields, poor management, and

TABLE 6.6
State Aid to Agriculture, 1950-80

Year	Share of Total State Aid to Agriculture in Total State Expenditures[a]	Share of State Investment in Agriculture in Total State Capital Investment
1950	3.1	11.4[d]
1951	3.7[b]	12.4[d]
1952	5.0	12.0[d]
1953	5.5[b]	13.8[d]
1954	5.6[b]	11.1[d]
1955	5.6[b]	10.9[d]; 7.8[e,f]
1956	7.5[b]	14.4[d]
1957	8.1	n.a.
1958	n.a.	10.5[f]
1959	4.1[c]	10.5[f]
1960	4.1[c]	13.0[f]
1963	16.0-18.0	n.a.
1964	16.0-18.0	n.a.
1978	12.7	10.7
1979	11.9	14.0
1970s (average)	10.0	n.a.
1980	12.6	17.8[g]

Sources:
Share of total state aid to agriculture: Jingji Yanjiu (Economic Research), No. 2, 1980, pp. 12-14 for 1950, 1957, 1963-64 & the 1970s average; derived from Nai-Ruenn Chen, Chinese Economic Statistics (Chicago: Aldine Publishing Co., 1967), p. 446 for 1951, 1953-56; derived from Audrey Donnithorne, China's Economic System (New York: Praeger, 1967), pp. 368, 383, 430 for 1959-60; derived from Beijing Review, No. 26, June 29, 1979, p. 10 & No. 39, September 29, 1980, pp. 11-12, 17 for 1978-79; Renmin Ribao, March 8, 1981, p. 3 for 1980.

Share of state investment in agriculture: derived from Chen, Chinese Economic Statistics, p. 446 for 1950-56; Yang Jianbai & Li Xuezeng, "The Relations Between Agriculture, Light Industry and Heavy Industry in China," Social Sciences in China, Vol. 1, No. 2 (June 1980), p. 190 for 1955 (1953-57 average), 1958-60; derived from Beijing Review, No. 29, July 20, 1979, pp. 17-18, 22; No. 26, June 29, 1979, p. 10 & No. 39, September 29, 1980, p. 17 for 1978-80.

n.a. = not available.

[a]Total state aid to agriculture includes state investment in agriculture and aid to agriculture for production and investment.

[b]Includes only state investment in agriculture.

[c]Average for 1959-60.

[d]Includes state investment in forestry.

[e]Average for 1953-57.

[f]Share of agriculture in total capital investment, including state investment and investment by the localities.

[g]Budgeted figure.

financial losses.[28]

Second, assistance to poor communes and brigades is a special item in the state budget established in 1959. Between 1959 and 1966, most of the funds were allocated to poor brigades for developing production, with a small share going to communes for developing industries. After 1967, the government shifted the priority in the use of the fund to agricultural mechanization. As a result, the collectives with better conditions received the bulk of the assistance.[29] In 1980, a separate fund for aid to economically backward regions was established, with 500 million yuan budgeted for 1980.[30] It is not clear what will happen to the fund for aid to poor communes and brigades.

Starting from 1981, the government has introduced some far-reaching fiscal reforms in the economy, which will have important implications for agriculture. To promote more efficient use of state funds, budgetary grants to state enterprises for investment will be changed to repayable bank loans.[31] It is not clear whether this applies to state farms and ranches. In any case, because the reforms will reduce state budgetary grants and increase state revenues, the government should be in a better position to extend more aid to rural areas.

In summary, through fiscal policies the government taxes agriculture on the the one hand and provides grants and investment to it on the other hand. In this way, the government channels part of the agricultural surplus that might otherwise be consumed into productive investment. We do not have sufficient data to calculate the net balance between the inflow of government funds and investment to agriculture and the outflow of taxes from agriculture. During the period 1953–71, state funds allocated to agriculture were 23.4 percent greater than the amount of agricultural tax collected.[32] The latter does not include rural commercial and industrial taxes, so the exact balance between the inflow and the outflow cannot be determined. The author's judgment is that there was a net outflow from agriculture in the early 1950s, and a net inflow to agriculture in the 1960s and 1970s. However, the net inflow was probably quite small, both as a percentage of total state expenditures and as a percentage of the state investment in the economy.

TABLE 6.7
Total State Aid and Bank Loans to Agriculture, 1950-80
(billion yuan)

Year	Total State Aid to Agriculture	Total Loans to Agriculture
1950-52	1.5[a]	1.6
1953-56	6.3[a]	6.3
1957-58	n.a.	6.1
1959-60	5.0	6.7[b]
1978	11.9	13.2
1979	15.4	17.0
1980	15.0	24.0

Sources: Chen, Chinese Economic Statistics, p. 446 for 1950-56 state aid and p. 469 for 1950-58 loans; Donnithorne, China's Economic System, pp. 368, 383 & 430 for 1959-60 state aid and p. 430 for 1953-60 loans; New China News Agency, January 23, 1980 for 1978-79; Renmin Ribao, March 8, 1981, p. 3 for 1980 state aid; Renmin Ribao, July 4, 1981, p. 2 for 1980 loans.

n.a. = not available.

[a]Includes state investment in agriculture only.

[b]Derived by subtracting the sum for 1953-58 (12.3 billion yuan) from the sum for 1953-60 (19 billion yuan).

Government Monetary Policies

Bank loans to agriculture are an important means of providing aid to agriculture. Most of the loans are made by the People's Bank, the Agricultural Bank of China, and the credit cooperatives, which are the banks' rural grass-roots agents at the commune level.[33] As shown in Table 6.7, the total amount of agricultural loans given in recent years is somewhat larger than the total amount of state expenditures on agriculture.

The interest rates charged by the People's Bank on various kinds of loans are given in Table 6.8. Before 1958, higher interest rates were charged on agricultural loans than on loans to state industrial and commercial enterprises; after 1958, the reverse was true. The interest rate differential between agricultural loans and loans to state industrial-commercial enterprises constitutes a state subsidy to agriculture.

Part of the agricultural loans comes from bank deposits made

TABLE 6.8
Monthly Rates of Interest on Loans Charged by State Banks,[a] 1953-81

Year	State Industrial Enterprises	State Commerical Enterprises	Agricultural Collectives
1953	0.45-0.48%	0.69%	0.75%
1955	0.48	0.60	0.60
1958	0.60	0.60	0.48
1959	0.60	0.60	0.60
1961	0.60	0.60	0.48
Before Sept. 1972	0.48	0.60	0.48
After Sept. 1972	0.42	0.42	0.36[b]
1981	0.20-0.30[c]	0.20-0.30[c]	n.a.

Sources: Pierre-Henri Cassou, "The Chinese Monetary System," China Quarterly, No. 59, July-September 1974, p. 562 for 1953-72; Renmin Ribao, November 27, 1980, p. 2 for 1981.

n.a. = not available.

[a]The rates for 1953-72 are charged by the People's Bank. Those for 1981 are charged by the Construction Bank of China.

[b]The rates of 0.18% for purchase of farm machinery and 0.36% for farm production (seeds, fertilizers, etc.) are given in Roland Berger, "Financial Aspects of Chinese Planning," Bulletin of Concerned Asian Scholars, Vol. 6 (1974), p. 17.

[c]For basic construction, the general rate is 0.25%. For coal mines, construction materials, postal and telecommunication units, the rate is 0.2%. For machinery, textile, crude-oil refining and petro-chemical enterprises, the rate is 0.3%.

by the rural population. In both 1978 and 1979, rural deposits amounted to about one-half of the agricultural loans. Thus, the banking system in China, as in other countries, helps to mobilize the agricultural surplus for investment in agriculture.

Interestingly, many agricultural loans are not repaid. For example, in Linfen County, Shanxi Province, more than 98 percent of the loans were not repaid in 1979.[34] In Kuandian County, Liaoning Province in 1980, "most of the agricultural loans cannot be repaid."[35] So many local banking units have arbitrarily forgiven the debts in violation of regulations that in 1979 the headquarters of the People's Bank of China issued a special directive, reminding the local units that the authority to forgive loans belongs to the central government and that the principle of "repayment of loans when due" must be thoroughly applied.[36]

The causes of the problem are varied and are related to the problems of the rural areas and of the banking system. The most important cause stems from the financial mismangement and difficulties of the collectives. In many cases, the financial difficulties are due to mismanagement by incompetent party cadres, including borrowing funds from the collectives for personal use thereby forcing the collectives to borrow from banks.[37] In other instances, the collectives purchased expensive and inappropriate farm machines under "strong sales pressures from some bureaus, and ended up in debt."[38]

Within the banking system, extreme leftist ideology has influenced local loan policies. During the Cultural Revolution decade, loan policies favored the poor collectives in spite of their inefficiency and poor repayment prospects. Loans were given, not for production purposes as intended, but for relief of poverty, which was the function of the state agricultural assistance fund. Also, the leftist ideology regarded giving loans to the richer collectives as capitalistic. Similarly, to demand the repayment of loans was equated with withdrawing support for agriculture.[39]

There are also practical problems in enforcing the repayment of loans. Some unemployment will be created if some collective enterprises in financial difficulties are required to make the repayment. Further, in the judicial system, there are no provisions for the courts, banks, or the public security agents to order sanctions against default.[40]

As a result of the default on loans, the amount of agricultural loans has been lower than what it otherwise would have been. Agricultural production is adversely affected because funds that should have been repaid and then reloaned to increase production were used for relief and consumption purposes.

Internal Sources of Investment

Internal sources of agricultural investment are those generated by and within agriculture itself. Three categories can be distinguished: Public accumulation funds, collective savings and investment in kind, and earnings of rural collective enterprises.

Public Accumulation Funds

Public accumulation funds are funds set aside from the collective's farming income primarily for investment purposes before the income is distributed to members of the collective.[41] This is the most important internal source of funds for agricultural investment. Obviously, the determinants of the accumulation funds are (1) the productivity and output of the collectives, (2) the prices of agricultural products and agricultural inputs, and taxes, which have already been discussed, and (3) the willingness of the peasants to save collectively in order to finance investment projects.

There are no aggregate data on the size of public accumulation funds. Fragmentary data on 17 collectives in 1971–77 indicate that these collectives saved 7–30 percent of their gross collective incomes, with an average of 16.4 percent saved.[42] This figure is very high for a country at a low level of per capita income and reflects the efficacy of China's rural collectives in controlling peasant consumption and promoting savings and investment.

Collective Savings and Investment in Kind

Collective savings and investment in kind, referred to by the Chinese as "labor accumulation," are the formation of capital projects primarily through the use of peasant labor in the collectives. Throughout the year, the collectives routinely undertake a number of capital projects, such as the reclamation of waste land and the construction of irrigation systems in their own areas. These are normally undertaken by members of the collectives during slack farming seasons, especially during the winter, so that the opportunity cost of the labor is relatively low. As the technology used is labor-intensive, little cash outlay occurs on the projects.

The peasants working on the projects are often awarded work points rather than cash wages. The work points entitle them subsequently to a share of the collective's income, leaving the other peasants with less to consume because the new projects will not immediately increase total output. Thus, in the final analysis, the investment is made possible by (1) the peasants working on the projects and forgoing what they would otherwise produce, and (2) other peasants supporting them through compulsory savings and reduced consumption. For this reason, the construction of the projects embodies the savings and investment in kind of the collectives. Naturally, when the projects are completed, the productive capacity of the collectives will be increased and all members of the collective will share in the increased output.

The Chinese concept of "labor accumulation" is similar to an idea popular in Western development economics. Nurkse has suggested that the underdeveloped countries mobilize their disguised unemployed rural labor to work on labor-intensive capital projects such as road construction. The resultant capital accumulation will have little opportunity cost because previously the workers were not productively employed.[43]

Collective savings and investment in kind are important in China since labor is abundant and since they enable the collectives, even the poor ones, to undertake relatively large, labor-intensive projects without relying on state assistance or on large cash income. In most underdeveloped countries, Nurkse's suggestion is difficult to implement because rural organizations for cooperative work are lacking and government initiatives are needed. Further, funds have to be appropriated to pay the peasant workers. In the case of China, the existing collective rural organizations and the system of rural income distribution make it possible for communities to undertake savings and investment in kind as a regular part of their agricultural investment activities. This is an example of the inherent advantages of collective farming in mobilizing rural human resources. Appropriately, the Chinese leaders have stressed that in rural capital construction, "we should mainly rely on accumulation of labor and not entirely on the state."[44]

Rural Collective Enterprises

The industrial and commercial enterprises owned and operated by the communes and brigades have become an important source of income in the rural economy. In the late 1950s, small rural plants were widely established as part of the Great Leap Forward. Although rural industries were deemphasized after the failure of the Great Leap, interest in them was revived since 1963 as part of the "walking on two legs" strategy of development. Both the collectives and the county governments were encouraged to develop rural industries to serve agriculture. As a result, rural industries expanded rapidly throughout most of the 1960s and 1970s. The collective enterprises consisted mainly of small plants or workshops for light industrial goods, handicrafts, processing of agricultural products, repair work, and various sideline production; the county-run enterprises tended to be small- or medium-size plants of the so-called five small industries: fertilizer, cement, farm machinery, energy (coal and hydropower), and iron and steel.[45] In 1979, there were 1.5 million rural enterprises.[46]

Since the late 1970s, with the new emphasis on agricultural modernization, the collective enterprises have been given an important role—to accumulate funds for agricultural modernization and to absorb the labor released by farm mechanization. On the other hand, the county-run enterprises, which dominated the five small industries, are to retrench and reorganize in order to increase the level of specialization and efficiency and to permit the larger modern plants to expand.

For the promotion of agricultural and rural development, the collective enterprises can play a very useful role, for many reasons. These enterprises do not require much capital to establish and, as their methods of production are labor-intensive, they can create many employment opportunities. For example, in 1978, more than 28 million peasants, or about 9.5 percent of the rural labor force, worked for the collective enterprises.[47] The collective enterprises provide many useful products and services to the rural areas, thereby reducing agriculture's dependence on the urban areas and relieving the strains on the transport system. Also, they usually utilize locally available raw materials, thus stimulating other local production, and they increase the level of rural income. Part of their earnings can be used to support agricultural investment. In 1978, the net profits of the rural collective enterprises were over 8 billion yuan, about twice the amount allocated by the state to aid the communes and brigades.[48] The enterprises invested 2.6 billion yuan in agriculture, which was more than 60 percent of the state's investment in basic construction in agriculture.[49]

In the operations of the collective enterprises, various problems have been encountered. Some are caused by the inefficiency of the enterprises themselves; others are caused by the local governments and party cadres. These problems can be put into four categories:

1. Production problems. Some plants are located where raw materials are not locally available, the skills of the workers are not high, or the scale of production is sometimes too low for efficiency. As a result, the quality of the products tends to be inferior. With proper planning and training, however, the collectives should be able to minimize these problems.

2. Burden of local taxes and charges. The industrial and commercial business taxes and income taxes that collective enterprises have to pay to the state have been mentioned previously in this chapter. In addition, local taxes and levies, which can be heavy and unreasonable, can be imposed. For example, in Liaoning Province in 1979,

> they [collective enterprises] are required to pay the local authorities 1.5 to 2 percent of their total sales as a contribution to the general fund for various projects and 1 to 1.5 percent for administrative expenses. They also have to take 35 to 40 percent out of their net profit after income tax and give it to the local authorities for general construction funds. In this way 80 percent of the income that staff and workers have earned with hard labor is taken. . . . Some neighbourhood offices appropriate vehicles of these enterprises for their own use, even though the enterprises still have to pay for the road tolls and petrol.[50]

From the point of view of economic incentives, it is inefficient and inconsistent for the state to subsidize the collectives for agricultural production and investment on the one hand, and to tax their enterprises heavily on the other hand, especially because the collective enterprises are being promoted to yield more funds for agricultural modernization. The cause of the problem is that local officials often act out of self-interest, in disregard of national interests and policies.

3. Periodic ideological restrictions. The periodic upsurge of extreme leftist ideology in the past has led to various unreasonable restrictions on the scope of enterprise operation. For example, many localities in the past have insisted on the principle of "three

locals" (local materials, local processing, and local sales) for the operation of collective enterprises, thus placing ideological restrictions on the scope of the enterprise operation. These restrictions were particularly severe and rigid during the Cultural Revolution. Thus, enterprises that did not directly serve the local areas were criticized as "grasping money rather than grasping the [party] line."[51] Some enterprises that had signed contracts with enterprises outside of local areas for subcontracting or sales were ordered by local party secretaries to cancel the contracts because they violated the principle.[52]

4. Discrimination against collective enterprises by state bureaucrats. The collective enterprises are discriminated against not only because ideologically and politically their status is inferior to that of the state enterprises, but also because they present potential competition to the state enterprises, which have long monopolized the local markets. Thus, the collectives tend to have difficulties in getting loans for their enterprises. They may experience difficulties in getting adequate fuel for transportation because their output is not part of the state plan, and thus their transportation needs are not taken into account by the local transportation department.[53] State stores or factories tend to be reluctant to purchase from collective enterprises because the latter compete with the state-run marketing units.[54] Nor can collectives sell products directly to consumers because it is illegal.[55] As late as 1980, when collective sideline production was being promoted to provide funds for agricultural modernization, charges of "blind expansion" were made against some collective agricultural processing plants because they competed with state processing plants.[56] In May 1981, new regulations on commune- and brigade-run enterprises were adopted by the State Council. The new regulations encourage the rural collectives to develop enterprises in many areas in accordance with local conditions, but order them to stop production in some areas (cotton textile, tobacco, and salt production). Also, where state enterprises have excess capacity in the processing of agricultural products, the collectives are not allowed to establish similar enterprises or to expand the capacity of existing ones.[57]

From the above discussion it is clear that the obstacles to the development of rural collective enterprises are not only economic but also political. The economic problems of production can be eliminated or reduced if careful planning and cost-benefit analysis are undertaken. The politically created problems are more difficult to overcome because they are embedded in the political-economic system itself—a system that permits the coexistence of the state sector and the collective sector (but with a second-class citizen status for the latter) and that places the collectives under the supervision and leadership of local party leaders. Although China's new leadership has reaffirmed the "self-management right" of the collectives and has urged the removal of undue restrictions on the collective enterprises, the latest regulations mentioned above still send conflicting signals to the rural collectives, reflecting the ambivalent attitude of the leadership toward such enterprises.

7
Summary, Evaluation, and Conclusions

Summary

Before evaluating China's agricultural performance and discussing the prospects, I will summarize the main points of the previous chapters.

China's natural resources for agriculture are relatively limited compared with the size of the population, which, by 1981, has exceeded one billion—nearly a quarter of the world's population. Only 11 percent of the land is cultivated, which constitutes only 7.3 percent of the world's total arable land. Because of topography and climate, water resources are very unevenly distributed regionally and seasonally, necessitating water-control works in both South and North China. Cropping patterns are conditioned by these factors as well. As possibilities of further expansion in cultivated land are very limited, future increases in agricultural output must come from increases in yield per unit of land.

Historically, early technological innovations and continuous expansion of cultivated area were the two sources of agricultural development in traditional China. Increases in agricultural output coupled with a relatively equal distribution of income in the country enabled the population to grow over the centuries. However, the low level of surplus over minimum consumption and the nonscientific orientation of the traditional Chinese culture inhibited the development of modern sciences. In the absence of modern sciences, the limits of field-experience-based agricultural technology were gradually reached by the nineteenth century. Although Western agricultural technology was introduced into China in the late nineteenth and early twentieth centuries, the efforts were too limited to have an appreciable impact on agricultural technology.

Thus, in 1949 the Chinese Communists inherited a poor, low-productivity agriculture. Under the leadership of Mao Zedong, they embarked on a series of institutional changes in the 1950s to socialize China's agriculture and to raise its productivity. Although the resultant collective system has proved effective in mobilizing rural resources, Mao's ideology and leadership also produced two periods of radical policies that were mostly counterproductive: the Great Leap Forward (1958-60) and the Cultural Revolution decade (1966-76). During these periods the peasants' incentives to produce were reduced by restrictions on rural markets, private plots, and family sideline production. Collective farming was also adversely affected by restrictive policies and incompetent local party leadership. During the Cultural Revolution decade, excessive emphasis on regional self-sufficiency in grain supply led to excessive land reclamation and deforestation with serious consequences for the environment.

There have also been two periods of supportive policies for agriculture: the early half of the 1960s and the post-Mao period since 1977. In both periods, the government adopted liberal policies to increase production by giving the peasants more material incentives. In particular, agriculture has been given a new priority in the drive to modernize the country.

As discussed in Chapters 4 and 5, significant though uneven progress has been made since the early 1950s in various areas of agricultural technology: fertilization, irrigation, mechanization, rural energy supply, plant breeding, and plant protection. In all these areas, the strategy of simultaneously developing modern technology in the state sector and labor-intensive or intermediate technology in the rural areas has contributed to making modern inputs (fertilizers, tools, hydropower) available to agriculture. The system of collective organization in the countryside has facilitated the diffusion of better technology, particularly in plant breeding and plant protection. It has also facilitated the mobilization of rural manpower for the construction of capital projects such as irrigation and land reclamation.

In some areas of labor-intensive technology—for example, local irrigation, organic fertilizers, small hydropower stations, biogas generation, and integrated pest control—China's achievement is impressive. In other areas, where a higher level of science and technology is needed—for example, plant breeding, chemical fertilizers, and farm machinery—China has not done as well. However, a high level of mechanization does not seem desirable, considering China's economic conditions.

As discussed in Chapter 6, since the early 1960s the government has gradually changed its financial policies (price, monetary, and fiscal policies) in favor of agriculture. However, the actual extent of government support given to agriculture through financial measures has not been very large. Price changes in favor of agriculture are offset to some extent by price violations and by the limited supply of subsidized industrial products. Government expenditures on agriculture are very limited in extent and are partially offset by the agricultural tax and the taxes on the collective enterprises. Thus, the collectives have had to rely primarily on their own resources for investment and development. Currently the government encourages the development of collective enterprises as a source of funds for agricultural modernization. Difficulties have been encountered by these enterprises, however, since they compete with the state-run enterprises.

Evaluation of Performance

In the previous chapters, we have examined the policies and programs for agricultural development and the concomitant institutional changes and technological progress in China's agriculture since 1949. The following questions naturally arise: What have been the results of these policies and programs in terms of output? How does China compare with other countries in agricultural performance?

Ability to Feed the Population

A useful point of departure for assessing China's agricultural performance is the important fact that with the exception of 1959–61, China has been able to feed a huge population without noticeable maldistribution and malnutrition. This is an important accomplishment that requires not only a reasonable productive agriculture, at the minimum, but also an equitable distribution system backed by an effective government.

Two factors have played an important role in making this possible: (1) the collective nature of China's agriculture, which has ensured work and income for all segments of the rural population, and (2) China's system of planned purchase and distribution of essential agricultural products, discussed in Chapter 6, which has ensured the availability of basic foodstuffs to all. China's achievement in feeding its population is impressive when compared with many developing countries, such as India, which have more favorable land-population ratios but cannot eradicate large pockets of food shortages and malnutrition.

Growth Rates

It is important to compare the growth rate of agricultural output over time with that of the population. In this respect, China's record varies from good to mediocre, depending on the time period chosen. The population growth rate has declined from about 2.3 percent in the mid-1950s to 2.1 percent in the 1960s and 1.6 percent in the 1970s.[1] For grain production, the average annual rate of growth from 1952 to 1979 was 2.7 percent; for cotton, it was 2 percent during the same period. During the early part of that period (1952–58) the average annual rate of growth was 4.2 percent for grain and 4.6 percent for cotton. From 1958, when the Great Leap Forward was launched, to 1978, it was only 2 percent and 1.3 percent for grain and cotton, respectively.[2] Thus, although China has managed to feed and clothe its population, the margin of surplus since 1958 has been, on the average, dangerously small. Imported grain has been needed to build up the reserves.

In Table 7.1, China's agricultural growth is compared with that of other developing countries in Asia. Taking the average output level of each country from 1967 to 1971 as 100, the level of total agricultural output for China was 118 in 1975 and 138 in 1980. These figures are slightly higher than the Asian averages and are moderately higher than those for Bangladesh, Burma, India, and Pakistan, but much lower than that of North Korea, South Korea, and Thailand. On a per capita basis, China's standing is

TABLE 7.1
International Comparison of Agricultural Production, 1975 and 1980
Production Indices[a]
(1969-71 = 100)

Country	Total Agricultural Production 1975	Total Agricultural Production 1980	Per Capita Agricultural Production 1975	Per Capita Agricultural Production 1980
China	118	138	109	119
Bangladesh	106	122	95	94
Burma	113	136	94	106
India	114	127	102	101
Indonesia	121	139	107	109
Japan	109	98	102	88
Korea, North	137	172	120	133
Korea, South	129	146	117	120
Malaysia	122	148	106	114
Pakistan	111	136	96	99
Philippines	128	155	111	115
Taiwan	104	111	95	92
Thailand	131	161	113	121
Asian Average	116	131	105	107
World Average	113	125	103	104

Sources: FAO Monthly Bulletin of Statistics, July-August 1981 for all countries except Taiwan; Taiwan Statistical Data Book, 1981.

[a]Indices of agricultural production are based on the sum of price-weighted quantities of different agricultural commodities produced after deductions of quantities used as seed and feed.

improved somewhat—ahead of Bangladesh, Burma, India, Indonesia, Pakistan, and the Philippines, but still behind North Korea and South Korea. The reason for the change is that the population growth rate has declined more rapidly in China than in the South and Southeast Asian countries.

The statistics for growth rates over a long period of time conceal the fact that the actual agricultural output in China has fluctuated in some years. As discussed in Chapter 3 and shown in Table 3.1, grain production declined substantially from 1959 to 1961, declined slightly in 1968, and did not grow at all from 1975 to 1977. Although these periods were associated with bad weather, some scientists have argued that short-sighted government policies during these periods either aggravated or were partially responsible for the results of bad weather. For example, the severity of the

TABLE 7.2
International Comparison of Rice and Wheat Yields Per Hectare, 1979
(kilograms per hectare)

Country	Paddy Rice	Wheat
China	3,717	1,500
Bangladesh	1,936	-
Burma	1,995	-
India	1,792	1,574
Indonesia	2,977	-
Japan	6,240	3,056
Korea, North	6,150	2,333[a]
Korea, South	6,556	2,118[a]
Malaysia	2,855	-
Pakistan	2,508	1,485
Philippines	2,000	-
Taiwan	4,459	2,580
Thailand	1,884	-
Asian Average	2,646	1,500
World Average	2,615	1,782

Sources: FAO Monthly Bulletin of Statistics, June 1980 and other issues; Taiwan Statistical Data Book, 1980.

[a]1978 figure.

1959-61 flooding has been attributed to the excessive irrigation campaign during the Great Leap Forward.[3] In recent years, various Chinese scientists have contended that excessive reclamation and deforestation have reduced annual precipitation over large areas, making droughts more frequent and severe.[4] The great flooding of 1981 in Sichuan has also been attributed by Chinese scientists to deforestation in the upper reaches of the Changjiang.[5]

Crop Yields

Another way to assess China's agricultural performance is to compare its yield per hectare of major crops with that of other countries. As shown in Table 7.2, China's rice yield of 3,717 kilograms per hectare in 1979 is above the Asian average of 2,646 kilograms. It is higher than that of the South and Southeast Asian countries but is far below that of Japan, North Korea, South Korea, and Taiwan. Because the East Asian countries have climatic conditions and work ethics closer to those of China, the latter comparison is more significant. With respect to wheat yield, China's 1,500 kilograms per hectare is not only far below that of its East Asian neighbors, but is no higher than the low yield level of the South Asian countries. The considerable disparity in rice and wheat yields between China on the one hand and Japan, North Korea, South Korea, and Taiwan on the other hand reflects China's relative lag in raising the productivity of its agriculture. Thus, in spite of the progress made in its agricultural technology, China still has a long way to go in catching up with its East Asian neighbors in agricultural productivity.

China's Own Assessment

How do the Chinese leaders assess their own agricultural performance? Not surprisingly, the answer varies with the time period and the leadership in question. During the Cultural Revolution decade, with the extreme leftists in charge of the mass media, China's agriculture was invariably described in glowing terms. After 1977, perhaps partly for political reasons, the assessment has been more realistic. For example, the fourth plenary session of the eleventh Central Committee of the Chinese Communist Party in 1979 described the agricultural growth in the previous thirty years as "sluggish," and attributed it to

> the absence of a stable social and political environment after the completion of socialist transformation; the adoption of some policies and measures which were unfavorable to arousing the peasants' enthusiasm for production; inadequate and ineffective state aid; neglect of technical innovation and agricultural research and education; and inefficient implementation of the principle of all-round development of farming, forestry, animal husbandry, side-line production and fishery.[6]

In 1981 in another official review of China's agricultural performance since 1949, Du Runsheng, a vice minister of the State Agricultural Commission, acknowledged that China had failed to eradicate dire poverty from large rural areas in spite of the ending of "exploitation." According to him, at the 1981 price levels, a peasant needs an annual income of 120 yuan to subsist. About one-third of the production teams in the country are above that level, making a per capita income of 120–1,000 yuan per year. Another third can reach that subsistence level "if they work hard." The final third of the production teams have an average annual per capita income of less than 60 yuan, and "they must rely on sideline occupations, on government relief or on frugality to make ends meet."[7]

The Political Factor

In my view, these official statements are candid and accurate assessments of China's agricultural performance and policy failings. It is precisely these policy failings that account for the fact that, in spite of the advantages of the rural collectives in mobilizing resources and promoting technological progress, China's agricultural growth rates and crop yields are only about average among the Asian nations.

What is not pointed out in the above statements is that, to a large extent, the nature of the political-economic system of China itself is responsible for these policy failings. Two aspects of the political-economic system should be mentioned in this connection. First, since China is a centrally planned economy under the leadership of the Chinese Communist Party, virtually all important aspects of agriculture—crop priorities, research emphasis, investment priorities, prices, agricultural taxes, and loans—are determined or profoundly affected by party policies, or more accurately by the views and personalities of the few top party leaders in power. Thus, changes in the leadership can and did have a profound unsettling effect on agriculture. This is the "absence of a stable social and political environment" quoted above.

Second, the existence of extensive party organizations in the rural areas and the power they have over the collectives often lead to abuses of power and arbitrary interference in farming activities, which nullify the advantages of collective farming. We have seen numerous examples of this in previous chapters.

The interference of local party leaders in the activities of the collectives is particularly undesirable when it affects decisions concerning agricultural technology because local party cadres often lack scientific and technical knowledge, as will be discussed later in this chapter, and are prone to make technological blunders. When these people are empowered to give "blind commands" concerning technical issues, the consequences can be disastrous. As we have seen in Chapter 4, during the Great Leap Forward many poorly designed irrigation projects without proper drainage were constructed at the insistence of local cadres, causing salinization of farmland and contributing to subsequent flooding. More recently, excessive digging of deep tube wells in Hepei in the 1970s has lowered the groundwater level and rendered useless almost half of the new wells. The excessive digging resulted from the provincial party committee's repeated refusal to heed the experts' warning of a possible "groundwater funnel" and the measures recommended to avoid it; the warning was dismissed as imaginary ("cannot possibly dig through to America") and as a scheme to oppose the "party line" and to discredit the provincial party committee.[8] Many other equally irrational technological "blind commands" have also been given by local party cadres in other areas of farming—for example, insisting on winter irrigation, which subsequently froze and destroyed the winter-wheat crop,[9] and insisting on multiple cropping where objective conditions were not appropriate, resulting in lower grain yield (see Chapter 5).

Conclusions

In light of the above, the verdict on China's agricultural performance cannot but be a mixed one. On the one hand, it has managed to feed a very large population equitably under difficult material conditions. The equitable aspect of the food distribution is a credit to its planning system, which is part and parcel of its political-economic system.

On the other hand, after more than three decades of large-scale and sometimes harsh "mass mobilization" of resources with the aid of the rural collective system, China's agricultural growth rates and crop yields are little more than average among the Asian countries and dire poverty still remains in many rural areas. The very political-economic system that has made "equity" possible in food distribution has also caused various types of "inefficiency" in the utilization of China's agricultural resources. This has caused China's agricultural performance to fall below its potential, a potential that should be relatively high among the developing countries because of China's effective rural collective organizations, improvement in agricultural technology, and hard-working peasants.

Problems and Prospects

What are the prospects for China's agriculture? Can China successfully modernize its agriculture by the year 2000, the target year for its Four Modernizations (agriculture, industry, science and technology, and defense)? These are difficult questions to answer because too many unknowns are involved. What we know with certainty is that, in addition to the problems already discussed above, a few potential and existing problems have to be forestalled or solved before agricultural modernization can be successful.

Future Political Instability

As discussed in Chapter 3, the post-Mao period has ushered in the latest swing of policy changes: peasant incentives are being increased; technological progress is given a top priority; state financial resources for agriculture are being increased; and the right of self-management by the collectives is reaffirmed. Coming after the policy excesses and failings of the Cultural Revolution decade, these changes are much needed and they augur well for China's agriculture. However, if the history of China since 1949 is any guide, further drastic political changes and policy reversals cannot be ruled out. There are simply too many high-level party, government, and military leaders who have lost power or feel threatened because their careers are based on the old radical politics. Further, there are many more cadres and officials at the middle and lower levels who cannot function adequately in the new modernization-oriented environment because of the political nature of their previous training and career advancement in an era of "red-over-expert" and "politics-in-command." Many of these people, no doubt, will welcome or even seek a return to the old order.

Thus, in the 1980s, this potential political discontent may pose the greatest danger to the current leadership and constitute the greatest obstacle to China's modernization. The best way to minimize this discontent and forestall radical political change, however, is not to replace these people en masse with new administrative and technical personnel. This will only heighten their insecurity and reduce their incentives to cooperate. Also there are simply not enough experienced administrators and competent technicians to replace them. In my view, the best strategy is to combine the new emphasis on technology and modernization and the refutation of the extreme leftist ideology and policies with increased scientific-technical education for these leaders and cadres.

Naturally, this will not be an easy task. Scientific-technical education has never been part of the training for party and government leaders and officials, and the average level of their scientific-technical knowledge is very low. For example, in 1979 the Ministry of Chemical Industry conducted its first training class for directors of small nitrogen-fertilizer plants. The trainees scored an average of 31 percent in their entrance examination—some did not even know the number of stages involved in nitrogen fertilizer production[10]—and this is by no means an isolated case. But precisely because the existing level of scientific-technical knowledge of most cadres and officials is very low, and given the fact that all of them cannot be replaced without serious consequences, government investment in their scientific-technical education will yield very high social returns.

Mechanization and Rural Unemployment

Ever since the 1950s, many Chinese leaders have believed that a modern agriculture is a highly mechanized one, characterized by a high level of labor productivity. As recently as late 1979, one of China's policy objectives was to "import, manufacture and popularize advanced farm machinery so as to raise labor productivity by a wide margin."[11] Although the policy of rapid mechanization has been modified since then, it remains the long-term objective of the Chinese leaders to raise agricultural labor productivity significantly by mechanization (see Chapter 5).

In an agriculture with abundant labor and limited land, what matters most is land productivity, that is, yield per unit of land, rather than labor productivity. As we have seen, China's grain yield is still far below that of its East Asian neighbors and there is still much scope for increasing total agricultural output by raising yield. On the other hand, efforts to raise labor productivity through the indiscriminate use of "advanced" farm machinery may not significantly raise total agricultural output, but will instead aggravate the problem of rural surplus labor, which in late 1981 amounted to as much as 30 percent of the rural labor force in some areas.[12] The urban areas also cannot provide employment to the displaced peasants because a large number of educated youths are already on a long waiting list for placement.[13] Furthermore, mechanization requires much fuel and electricity, which the country can ill afford. Thus, agricultural mechanization should be more gradual and selective, and part of the resources allocated to mechanization could be better spent on chemical fertilizers to raise yield or on improving the transportation network.

Transportation Bottleneck

Current policies correctly emphasize the development of a diversified agriculture, with regions specializing in activities, such as grain production, nongrain farming, animal husbandry, forestry, and fishery, in which they have a comparative advantage. It should be realized, however, that a greater degree of specialization in agricultural production will necessitate more interregional exchanges of agricultural products and will, therefore, require expansion in storage facilities and long-distance transport capabilities to accommodate the exchanges. Similarly, increases in the use of chemical fertilizers, commercial seeds, and farm machinery will also require more storage facilities and transportation.

Even at the current low level of agricultural specialization and interregional exchanges, refrigeration and storage facilities are highly inadequate, forcing state purchasing stations at times to refuse purchase of products delivered to the stations (see Chapter 6). Transportation facilities are even more heavily strained, constituting probably the greatest bottleneck in the economy. As shown in Table 7.3, the nation had less than 900,000 kilometers of roads in 1979, or less than 100 kilometers per 1,000 square kilometers of land. The railway network is in a worse situation. Between 1952 and 1978, the length of rail track merely doubled, whereas rail freight traffic increased by a factor of eight, resulting in extreme congestion and inefficiency. In terms of railway density, China has about 5.4 kilometers of rail track per 1,000 square

TABLE 7.3
Transport Development, 1949-80

Year	Railway Length (thousand km)	Road Network (thousand km)	Railway Freight Transport (billion ton-km)
1949	22.0	80.7	18.4
1952	24.5	126.7	60.2
1957	29.9	254.6	134.6
1960	33.0	500.0	228.0
1965	37.4	514.5	269.6
1970	40.0	650.0	298.0
1973	n.a.	726.9	403.0
1975	48.4	783.6	424.6
1977	49.1	855.0	455.8
1978	50.0	890.0[a]	533.3
1979	51.5	875.8	558.0
1980	n.a.	n.a.	571.7

Sources: Annual Economic Report of China, 1981, part VI, p. 19 for 1949-57, 1965, 1975, 1979; Christopher Howe, China's Economy (New York: Basic Books, 1978), p. 122 for 1960, 1970, 1973; Beijing Review, No. 27, July 6, 1979, p. 39 for 1977-78; Beijing Review, No. 20, May 19, 1980, p. 20 & Far Eastern Economic Review Asia Yearbook, 1981, p. 127 for 1979; Beijing Review, No. 19, May 11, 1981, p. 27 for 1980.

n.a. = not available.

[a]This figure is probably inaccurate in the light of the 1979 figure.

kilometers of land, which is very low even by Asian standards (see Table 7.4).

The expansion of storage facilities and the transportation system will require much investment, thus competing with other sectors of the economy for limited state funds. Yet it is essential that they keep pace with agricultural growth and diversification if the objective of an efficient agriculture is to be attained.

Conclusions

The obstacles to China's agricultural modernization by the year 2000 are many and enormous. However, they are not insurmountable; they are amenable to government actions and thus can be overcome through appropriate policies and effective measures. In the final analysis, in China's political-economic

TABLE 7.4
International Comparison of Railway Density, 1980

Country	Railway Length (kilometers)	Land Area (1000 sq. km)	Railway Density (km/1000 sq. km)
China	51,500[a]	9,561[b]	5.4
Bangladesh	2,874	144	20.0
Burma	2,705	677	4.0
India	61,271	3,288	18.6
Indonesia	6,637	1,904	3.5
Japan	35,143	378	93.0
Korea, North	4,322[a]	122	35.4
Korea, South	5,860	99	59.2
Malaysia	1,666	330	5.0
Pakistan	8,915	796	11.2
Philippines	3,503	297	11.8
Taiwan	1,086	36	30.2
Thailand	3,825	513	7.5

Sources: Far Eastern Economic Review Asia Yearbook, 1981; Table 7.3.

[a]1979 figure.

[b]Excluding Taiwan.

system, it is party policies, and state resources to back the policies, that are most important in agricultural and economic development.

Assuming that political stability can be ensured and the current agriculture-first strategy of development is continued, the prospects for China's agricultural development appear auspicious. Under these favorable conditions, the most important constraint on the pace of agricultural development will be the total amount of resources that China will have for its Four Modernizations and the share that the Chinese leaders will allocate to agriculture. As the recent cutbacks in investment projects demonstrated, it is unlikely that China will have sufficient resources to help propel simultaneously all Four Modernizations to the ambitious "advanced world level" by the year 2000. Thus, painful choices will have to be made in the allocation of resources among sectors of the economy and in the scaling down of the grandiose plans.

In making the difficult choices to allocate the resources, it is imperative that agriculture be given the highest priority, for it is only then that China's agriculture can be modernized and become the "foundation" of the economy in reality as well as in theory. A highly efficient agriculture will not only raise the standard of living of the large rural population, but will also support the rest of the economy by providing it with abundant foodstuffs, raw materials, investable surplus, skilled manpower, and a vast market for industrial goods. The modernization of industry cannot but result.

Notes

Chapter 1

1. Unless otherwise indicated, basic data given in this chapter come from: Qi Wen, *China: A General Survey* (Beijing: Foreign Languages Press, 1979) and Chung Chih, *An Outline of Chinese Geography* (Beijing: Foreign Languages Press, 1978).

2. China's cultivated area is estimated at about 100 million hectares in the late 1970s. For details, see Chapter 4. The world's total arable land in 1978 is estimated at 1,462 million hectares in *FAO Production Yearbook, 1978* (Rome: Food and Agricultural Organization of the United Nations, 1979).

3. *China Reconstructs*, March 1980, p. 5: *Beijing Review*, no. 17, April 28, 1980, p. 28.

4. *New York Times*, April 7, 1980, p. 12.

5. *Ta Kung Pao*, September 20, 1979, p. 18.

6. This classification follows that of John L. Buck, *Land Utilization in China* (Chicago: University of Chicago Press, 1937), pp. 131–132. See also T. H. Shen, *Agricultural Resources of China* (Ithaca: Cornell University Press, 1951), pp. 18–29; T. R. Tregear, *China: A Geographical Survey* (New York: Halstead Press, 1980), pp. 27–32.

7. Qi, *China*, p. 91.

8. *China Reconstructs*, August 1979, p. 24.

9. *Peking Review*, no. 26, June 30, 1978, p. 10.

10. *Ta Kung Pao*, September 27, 1979, p. 3, gives 33 million hectares; *Hongqi*, no. 23, 1980, p. 41, gives 53 million hectares.

11. *Peking Review*, no. 26, June 30, 1978, p. 10.

12. *Renmin Ribao*, November 25, 1979, editorial.

13. *Beijing Review*, no. 13, March 31, 1980, p. 5. The figure for the United States and Europe is about 30 percent. China ranks 120th in the world in the percentage of land covered with trees.

14. *Guangming Ribao*, March 11, 1980, p. 1.

15. *Renmin Ribao*, October 10, 1978, p. 2.

16. Silting in the Changjiang due to soil erosion in the upper reaches has increased so much that some experts fear that it might become the second "yellow" river in China. See *Guangming Ribao*, December 22, 1979, and September 25, 1980, p. 2.

17. *Beijing Review*, no. 13, March 31, 1980, p. 5.

18. *China Reconstructs*, August 1977, p. 43.

19. *Guangming Ribao*, October 9, 1980, p. 1.

20. The classification of agricultural regions and areas follows primarily that of Buck, *Land Utilization*, pp. 23–91. See also Tregear, *China*, pp. 130–182.

Chapter 2

1. See, for example, Arnold J. Toynbee, *A Study of History*, 10 vols. (London: Oxford University Press, 1934–54), vol. 1, pp. 318–321.

2. Karl A. Wittfogel, *Oriental Despotism: A Comparative Study of Total Power* (New Haven, Conn.: Yale University Press, 1957).

3. Ping-ti Ho, "The Loess and the Origin of Chinese Agriculture," *American Historical Review*, vol. 75 (1969), pp. 1–14; Ping-ti Ho, *The Cradle of the East* (Hong Kong: The Chinese University, 1975), chap. 2.

4. Ho, "Loess and the Origin," pp. 6, 12, 15.

5. Ibid., pp. 26–27.

6. Ibid., pp. 19–26.

7. Cho-yuan Hsu, *Han Agriculture: The Formation of Early Chinese Agrarian Economy* (Seattle: University of Washington Press, 1980), pp. 119–121.

8. Ibid., p. 121.

9. Qi Wen, *China: A General Survey* (Beijing: Foreign Languages Press, 1979), p. 81.

10. Hsu, *Han Agriculture*, pp. 123–124.

11. For a concise survey of the encyclopedia, see Shih Sheng-han, *A Preliminary Survey of the Book Ch'i Min Yao Shu* (Beijing: Science Press, 1974), 2nd edition.

12. Mark Elvin, *The Pattern of the Chinese Past* (Stanford, Calif.: Stanford University Press, 1973), p. 129.

13. Ibid., pp. 118–119.

14. Evelyn S. Rawski, *Agricultural Change and the Peasant Economy of South China* (Cambridge, Mass.: Harvard University Press, 1972), p. 41.

15. Elvin, *Chinese Past*, p. 121.

16. Ping-ti Ho, *Studies on the Population of China, 1368–1953* (Cambridge, Mass.: Harvard University Press, 1959), pp. 183–184.

17. Elvin, *Chinese Past*, p. 203.

18. Albert Feuerwerker, "The Chinese Economy, ca. 1870–1911" (Ann Arbor: University of Michigan Papers in Chinese Studies, No. 5, 1969), p. 16.

19. Dwight H. Perkins, *Agricultural Development in China, 1368–1968* (Chicago: Aldine Publishing Co., 1969), pp. 13–17.

20. Feuerwerker, "Chinese Economy," p. 6.

21. Chi-ming Hou and Tzong-shian Yu (eds.), *Modern Chinese Economic History* (Taipei: Institute of Economics, Academia Sinica, 1979), pp. 140, 160–161.

22. Ibid., pp. 140–143.

23. Ibid., p. 149.

24. Feuerwerker, "Chinese Economy," p. 66.

25. Hou and Yu, *Economic History,* pp. 144–148.

26. Albert Feuerwerker, "Economic Trends in the Republic of China, 1912–1949" (Ann Arbor: University of Michigan Papers in Chinese Studies, no. 31, 1977), p. 44.

27. Cited in Feuerwerker, "Chinese Economy," p. 6.

28. T. H. Shen, *Agricultural Resources of China* (Ithaca, N.Y.: Cornell University Press, 1951), p. 355; T. H. Shen, "First Attempts to Transform Chinese Agriculture, 1927–1937," in Paul K. T. Sih (ed.), *The Strenuous Decade: China's Nation-Building Efforts, 1927–1937* (New York: St. John's University, 1970), pp. 210–232.

29. Franklin L. Ho's comments in Sih, *Strenuous Decade,* p. 235.

30. Feuerwerker, "Economic Trends," pp. 82–83.

31. Feuerwerker, "Chinese Economy," p. 16.

32. Feuerwerker, "Economic Trends," p. 49.

33. Ibid., p. 48.

34. John L. Buck, *Land Utilization in China* (Chicago: University of Chicago Press, 1937), p. 196.

35. Hou and Yu, *Economic History,* p. 165.

36. Feuerwerker, "Economic Trends," p. 63.

37. Cited in Ho, *Population of China,* p. 223.

38. R. H. Tawney, *Land and Labor in China* (Boston: Beacon Press, 1966), pp. 72, 77.

39. Ibid., p. 11.

40. Compare Riskin's summary of the debate between the distributional and the technological school of thought in D. H. Perkins (ed.), *China's Modern Economy in Historical Perspective* (Stanford, Calif.: Stanford University Press, 1975), pp. 56–64. See also Philip C. C. Huang (ed.), *The Development of Underdevelopment in China* (White Plains, N.Y.: M.E. Sharpe, 1982), pp. 1–73, for an alternative view that regards class structure as the primary cause of underdevelopment in traditional China.

41. Elvin, *Chinese Past*, pp. 204, 230.

42. Anthony M. Tang, "China's Agricultural Legacy," *Economic Development and Cultural Change*, vol. 28 (1979), p. 10.

43. Feuerwerker, "Chinese Economy," p. 14; Rawski, *Agricultural Change*, pp. 29-30, 144-146.

Chapter 3

1. For more detailed discussion of the role of agriculture in economic development, see Bruce F. Johnston and John W. Mellor, "The Role of Agriculture in Economic Development," *American Economic Review*, vol. 51 (September 1961), pp. 566-593.

2. See, for example, Liu Shaoqi's 1950 speech on agrarian reform in Mark Selden (ed.), *The People's Republic of China: A Documentary History of Revolutionary Change* (New York: Monthly Review Press, 1979), pp. 236-240.

3. For vivid accounts of the process of land reform in some villages, see William Hinton, *Fanshen: A Documentary of Revolution in a Chinese Village* (New York: Vintage Books, 1966); Jack Belden, *China Shakes the World* (New York: Monthly Review Press, 1970).

4. The per capita landholding after the land reform was about 2 to 3 mou in Henan province, 1 to 2.5 mou in Henan and Hebei provinces, and 1.5 to 2 mou in East China. One mou is about 0.067 hectare. See Chao Kuo-Chün, *Agrarian Policies of Mainland China: A Documentary Study, 1945-1965* (Cambridge, Mass.: Harvard University Press, 1957), p. 37. For a detailed technical analysis of the impact of the land reform, see Victor D. Lippit, *Land Reform and Economic Development in China* (White Plains, N.Y.: International Arts and Science Press, 1974).

5. Yang Jianbai and Li Xuezeng, "The Relations Between Agriculture, Light Industry and Heavy Industry in China," *Social Sciences in China*, vol. 1, no. 2 (June 1980), p. 190. Total investment includes investment by both the state and the localities.

6. See Mao's 1955 speech, "On the Question of Agricultural Cooperation" in Selden, *People's Republic*, pp. 342-350.

7. Kenneth Walker, *Planning in Chinese Agriculture: Socialisation of the Private Sector, 1956-1962* (Chicago: Aldine Publishing Co., 1965), pp. 61-62.

8. Three major reasons for this difference can be given. (1) Many Chinese peasants had acquired their land free only a few years previously, during the 1949 Land Reform, but the Russian peasants had to pay dearly for their land in the nineteenth century, long before the Soviet collectivization of 1929. (2) The Chinese Communists were more experienced with the peasants; the Bolsheviks were more urban based.

(3) China received Soviet aid in the First Five-Year Plan, which lessened the need for heavy taxation of agriculture to finance industrial development, but that was not the case with Soviet collectivization. This final point is discussed by Gregory Grossman in Donald Treadgold (ed.), *Soviet and Chinese Communism* (Seattle: University of Washington Press, 1967), p. 294.

9. K. C. Yeh, "Soviet and Chinese Industrialization Strategies," in Treadgold, *Soviet and Chinese Communism*, p. 345, footnote 54.

10. Ibid., p. 355.

11. Cited in Choh-ming Li (ed.), *Industrial Development in Communist China* (New York: Praeger, 1964), p. 22.

12. Kang Chao, *Agricultural Production in Communist China, 1949–1965* (Madison: University of Wisconsin Press, 1970), pp. 132–134.

13. China has been exporting an average of about 1 million metric tons of rice every year so that the net import of grain has been about 3 to 4 million metric tons a year. For details see Frederic M. Surls, "China's Grain Trade," in Joint Economic Committee, *Chinese Economy Post-Mao* (Washington, D.C.: U.S. Government Printing Office, 1978), pp. 653–670.

14. Benjamin Schwartz, "China's Developmental Experience, 1949–72," in Michel Oksenberg (ed.), *China's Developmental Experience* (New York: Praeger, 1973), p. 24.

15. Alva L. Erisman, "China: Agricultural Development, 1949–71," in Joint Economic Committee, *People's Republic of China: An Economic Assessment* (Washington, D.C.: U.S. Government Printing Office, 1972), p. 133.

16. For a detailed discussion of the pros and cons of mass science during the Cultural Revolution decade, see Robert C. Hsu, "Mass Science in China," *Bulletin of the Atomic Scientists*, vol. 35, no. 2 (February 1979), pp. 27–30.

17. Chengdu, Sichuan provincial service, January 28, 1980.

18. *Renmin Ribao*, July 24, 1978, p. 2.

19. *Renmin Ribao*, December 2, 1978, p. 1; January 18, 1979, p. 2, June 25, 1979, p. 2; Lanzhou, Gansu provincial service, October 11, 1979.

20. *Beijing Review*, no. 12, March 23, 1979, p. 13.

21. *Beijing Review*, no. 26, June 29, 1979, p. 14.

22. *China Reconstructs*, July 1979, p. 6.

23. *Fei Qing Yue Bao* [Monthly bulletin on the Communist conditions], Taipei, vol. 23, no. 4 (1980), p. 11.

24. *China Reconstructs*, March 1981, p. 9; *Renmin Ribao*, April 8, 1980, p. 3; April 15, 1979, p. 2.

25. *Beijing Review*, no. 26, June 29, 1979, p. 10.

26. *Beijing Review*, no. 45, November 9, 1979, p. 6.
27. *China Reconstructs*, March 1981, p. 9.

Chapter 4

1. The state farms are run by municipal or provincial authorities as state enterprises. However, they constitute only a small part of agriculture. In 1980, China had some 2,000 state farms with 4.2 million hectares of farmland, about 4 percent of the national total, employing 4.8 million people. See *Beijing Review*, no. 31, August 4, 1980, p. 17.
2. Dwight H. Perkins, *Agricultural Development in China, 1368–1968* (Chicago: Aldine Publishing Co., 1969), pp. 70–71.
3. Ibid., p. 71.
4. John L. Buck, *Land Utilization in China* (Chicago: University of Chicago Press, 1937), p. 259.
5. Kang Chao, *Agricultural Production in Communist China, 1949–1965* (Madison: University of Wisconsin Press, 1970), p. 147.
6. *Renmin Ribao*, January 29, 1959, p. 1, cited in Leslie T. C. Kuo, *Agriculture in the People's Republic of China: Structural Changes and Technical Transformation* (New York: Praeger, 1976), p. 142.
7. Calculated from Tables 4.1 and 4.2.
8. Kuo, *Agriculture*, p. 144.
9. *Renmin Ribao*, January 25, 1980, p. 2.
10. For details of the production technology used in some small chemical-fertilizer plants, see American Rural Industry Delegation, *Rural Small-Scale Industry in the People's Republic of China* (Berkeley: University of California Press, 1977), pp. 154–176.
11. Kang Chao, "The Production and Application of Chemical Fertilizers in China," *China Quarterly*, no. 64 (December 1975), p. 713.
12. *Renmin Ribao*, January 15, 1980, p. 3.
13. The supply and marketing cooperatives were established in 1950–51 in the countryside to replace private merchants for rural commercial activities. Since then they have become the state's grassroot agents in the rural areas for sales of agricultural inputs and consumer goods and the procurement of local sideline products. State grain procurement, however, is made through a separate network of grain purchasing stations. For the activities and organization of the supply and marketing cooperatives, see *China Reconstructs*, November 1980, pp. 10–11; Jung-chao Liu, *China's Fertilizer Economy* (Chicago: Aldine Publishing Co., 1970), pp. 74–82.
14. Liu, *China's Fertilizer Economy*, pp. 84–94
15. Ibid., p. 74.

16. Chao, *Agricultural Production*, p. 162.
17. *Renmin Ribao*, April 23, 1980, p. 2.
18. *Renmin Ribao*, August 4, 1980, p. 2.
19. *Renmin Ribao,* December 25, 1962, p. 5, cited in Kuo, *Agriculture,* p. 148.
20. *Guangming Ribao*, June 5, 1980, p. 2.
21. *Renmin Ribao,* January 24, 1981, p. 2.
22. *Guangming Ribao,* June 5, 1980, p. 2, and *FAO Fertilizer Yearbook, 1980,* pp. 34–36, 43. The former gives 1:0.2:0.004 as China's ratio, but a comparison with FAO figures suggests that 1:0.2:0.04 is more accurate.
23. *Renmin Ribao*, March 12, 1980, p. 3; March 31, 1980, p. 4; April 23, 1980; July 28, 1980, p. 2.
24. Kuo, *Agriculture*, p. 125; Chao, *Agricultural Production*, p. 120.
25. Perkins, *Agricultural Development*, pp. 61–63.
26. *China Reconstructs*, December 1979, p. 31.
27. The "Eight-Character Charter of Agriculture" refers to the eight technical measures in agriculture that Mao considered as the most important in raising agricultural yields per unit of land: water conservation, fertilization, soil conservation, plant breeding, close planting, plant protection, tool improvement, and field management.
28. Chao, *Agricultural Production*, p. 123.
29. Kuo, *Agriculture*, pp. 116–117.
30. Chao, *Agricultural Production*, pp. 130–133.
31. Thomas B. Wiens, "The Evolution of Policy and Capabilities in China's Agricultural Technology," in Joint Economic Committee, *Chinese Economy Post-Mao* (Washington, D.C.: U.S. Government Printing Office, 1978), p. 692.
32. *Renmin Ribao*, September 18, 1980, p. 2.
33. *Renmin Ribao*, August 13, 1979, p. 2.
34. *Guangming Ribao*, March 18, 1980, p. 2.
35. *Beijing Review*, no. 19, May 11, 1979, pp. 6–7.
36. *Renmin Ribao*, January 31, 1980, p. 1.
37. *Renmin Ribao*, January 28, 1980, p. 4.
38. The American Plant Studies Delegation, *Plant Studies in the People's Republic of China* (Washington, D.C.: National Academy of Sciences, 1975), p. 25.
39. Wiens, "Evolution of Policy," pp. 676–677.
40. Ibid., p. 677.
41. *China Reconstructs*, May 1980, p. 44.
42. Ibid.
43. *China Reconstructs*, November 1977, p. 37.

44. New China News Agency, February 11, 1980.
45. American Plant Studies Delegation, *Plant Studies*, p. 26.
46. Wiens, "Evolution of Policy," p. 677.
47. *Renmin Ribao*, March 23, 1980, p. 2; *Beijing Review*, April 14, 1980, p. 27.
48. *Beijing Review*, April 14, 1980, p. 27.
49. American Plant Studies Delegation, *Plant Studies*, p. 26.
50. Kuo, *Agriculture*, p. 187.
51. *China Reconstructs*, June 1980, p. 29.
52. *Renmin Ribao*, April 1, 1981, p. 2.
53. New China News Agency, Fuzhou, July 12, 1976.
54. Food and Agriculture Organization, *Learning from China: A Report on Agriculture and the Chinese People's Communes* (Rome: FAO, 1978), p. 24; Wiens, "Evolution of Policy," p. 680.
55. American Plant Studies Delegation, *Plant Studies*, p. 147.
56. *Remin Ribao*, November 30, 1979, p. 2; September 10, 1980, p. 5.
57. American Plant Studies Delegation, *Plant Studies*, p. 147.
58. *Renmin Ribao*, December 8, 1980, p. 2.
59. *Renmin Ribao*, December 24, 1981, p. 2.

Chapter 5

1. See Robert C. Hsu, "Agricultural Mechanization in China: Policies, Problems, and Prospects," *Asian Survey*, vol. 19, no. 5 (May 1979), p. 437 for detailed references.
2. Ibid.; Kang Chao, *Agricultural Production in Communist China, 1949–1965* (Madison: University of Wisconsin Press, 1970), pp. 92–93.
3. Chao, ibid., p. 95.
4. Leslie T.C. Kuo, *Agriculture in the People's Republic of China: Structural Changes and Technical Transformation* (New York: Praeger, 1976), pp. 101–103.
5. For details, see Benedict Stavis, *The Politicis of Agricultural Mechanization in China* (Ithaca, N.Y.: Cornell University Press, 1978), pp. 126–162.
6. Ibid., pp. 93–97.
7. Kuo, *Agriculture,* p. 246.
8. Thomas B. Wiens, "Evolution of Policy and Capabilities in China's Agricultural Technology," in Joint Economic Committee, *China's Economy Post-Mao* (Washington, D.C.: U.S. Government Printing Office, 1978), p. 696.
9. New China News Agency, October 21, 1968.
10. *Renmin Ribao*, June 17, 1977, p. 3; *Peking Review*, no. 8, February 24, 1978, p. 11.

11. *Peking Review*, no. 7, February 17, 1978, p. 9; no. 23, June 9, 1978, p. 6.

12. Hsu, "Agricultural Mechanization," pp. 445–446. In early 1980, China had 1,900 farm machinery plants, most of them small ones. See *Beijing Review*, no. 5, February 4, 1980, p. 7.

13. *Renmin Ribao*, April 17, 1979, p. 2.

14. *Renmin Ribao*, February 6, 1979, p. 1.

15. New China News Agency, October 8, 1979.

16. New China News Agency, November 20, 1979.

17. *China Reconstructs*, August 1981, p. 54.

18. Ibid.

19. *Beijing Review*, no. 41, October 12, 1981, p. 3.

20. The reluctance to mechanize because of high cost or labor displacement effect is best documented in several articles in *Guangming Ribao*, September 6, 1980, p. 4, and *Renmin Ribao*, October 27, 1980, p. 2.

21. *Renmin Ribao*, December 16, 1978, p. 3. The possibility of lower total output from triple cropping has also been raised in Wiens, "Evolution of Policy," p. 701.

22. *Renmin Ribao*, October 19, 1980, p. 3.

23. It has been complained that the collectives often had to use up all of their savings in order to purchase the machines. See, for example, *Guangming Ribao*, September 6, 1980, p. 4.

24. For example, in a county in Shanxi Province it has been estimated that the cost of full mechanization for the whole county will be 500 million yuan. The average annual state subsidy for mechanization to the county between 1974 and 1978 was 1.3 million yuan. See *Renmin Ribao*, May 15, 1979, p. 2.

25. *Renmin Ribao,* July 13, 1979, p. 2.

26. *Peking Review,* no. 23, June 9, 1978, p. 5.

27. *Hungqi,* no. 4, February 16, 1980, p. 2. However, according to *Annual Economic Report of China, 1981* (Beijing: Academy of Social Sciences, 1981), part 6, p. 13, in 1975 and 1979, 33.3 and 42.4 percent of the farmland were plowed by tractors, respectively.

28. *China Reconstructs*, August 1981, pp. 53–54.

29. *Beijing Review*, no. 36, September 7, 1981, p. 5; *Renmin Ribao*, July 23, 1981, p. 1; July 26, 1981, p. 5; December 24, 1981, p. 2.

30. American Rural Industry Delegation, *Rural Small-Scale Industry in the People's Republic of China* (Berkeley: University of California Press, 1977), p. 153.

31. *Beijing Review*, no. 35, September 1, 1980, p. 4; *Renmin Ribao*, October 27, 1980, p. 5.

32. *Guangming Ribao*, July 19, 1980, p. 4.

33. Ibid.

34. Although some local areas attempted to achieve self-sufficiency in firewood in the 1960s during the Cultural Revolution, no effective measures were taken. Symbolic of the political orientation of the Cultural Revolution, a provincial symposium on the provision of firewood held in Guangdong Province in 1968 concluded that "the key to success in the production of firewood and thrift in the use of coal lay in the study and application of Mao Zedong's thought in a living way." Guangzhou, Guangdong provincial service, May 23, 1968.

35. Chao, *Agricultural Production*, p. 140.

36. Kuo, *Agriculture*, p. 241.

37. Chao, *Agricultural Production*, p. 140.

38. Kuo, *Agriculture*, p. 242.

39. *Renmin Ribao*, November 6, 1980, p. 2.

40. *Renmin Ribao*, August 6, 1980, p. 1.

41. Ibid.

42. *Renmin Ribao*, January 18, 1980, p. 1.

43. *Renmin Ribao*, October 24, 1980, p. 1.

44. American Rural Industry Delegation, *Rural Small-Scale Industry*, p. 134.

45. *Renmin Ribao*, March 17, 1980, p. 3.

46. *Renmin Ribao*, December 30, 1980, p. 1.

47. *Renmin Ribao*, July 8, 1980, p. 5. A Western estimate puts these figures at 540 and 300 million kilowatts, respectively. See Vaclav Smil, "China's Energetics: A System Analysis," in Joint Economic Committee, *Chinese Economy Post-Mao*, p. 348.

48. *Renmin Ribao*, February 28, 1980, p. 5.

49. *Renmin Ribao*, March 6, 1980, p. 2.

50. *China Reconstructs*, July 1977.

51. *Renmin Ribao*, October 20, 1981, p. 4; *Annual Economic Report of China, 1981*, part 4, pp. 60–61.

52. *Renmin Ribao*, July 8, 1980, p. 5.

53. *Peking Review*, February 24, 1978, p. 6.

54. *Annual Economic Report of China, 1981*, part 4, pp. 64–65.

55. *Guangming Ribao*, July 19, 1980, p. 4.

56. Bobby A. Williams, "The Chinese Petroleum Industry: Growth and Prospects," in Joint Economic Committee, *China: A Reassessment of the Economy* (Washington, D.C.: U.S. Government Printing Office, 1975), pp. 231–232.

57. *New York Times*, January 19, 1982.

58. Smil, "China's Energetics," p. 347.

59. *Renimin Ribao*, July 20, 1980, p. 1.

60. *Beijing Review,* no. 35, September 1, 1980, p. 30.

61. As testimonies to China's world leadership in biogas utilization, the United Nations Environmental Program sponsored a month-long study project in 1979 in China to study China's biogas technology. In 1980, the United Nations Industrial Development Organization sponsored a conference on methane in China for specialists from 21 developing countries.

62. *Guangming Ribao,* July 13, 1980, p. 3.

63. *Guangming Ribao,* July 19, 1980, p. 4.

64. *Guangming Ribao,* July 18, 1980, p. 4.

65. *Renmin Ribao,* May 9, 1980, p. 2.

66. *Guangming Ribao,* July 19, 1980, p. 4.

67. *Encyclopaedia Britannica* (1974), Macropaedia, vol. 14, p. 139.

68. Kuo, *Agriculture,* p. 207.

69. *China Reconstructs,* June 1978, p. 35.

70. Ibid.

71. *Peking Review,* no. 37, September 15, 1978, p. 14.

72. Ibid., p. 15.

73. The American Plant Studies Delegation, *Plant Studies in the People's Republic of China* (Washington, D.C.: National Academy of Sciences, 1975), p. 144.

74. *China Reconstructs,* June 1978, p. 36.

75. American Insect Control Delegation, *Insect Control in the People's Republic of China* (Washington, D.C.: National Academy of Sciences, 1977), pp. 75–79.

76. Ibid., pp. 76–77.

77. Ibid., p. 92.

78. Boel Berner, "The Organization and Economy of Pest Control in China," (Lund, Sweden: Research Policy Institute, Discussion Paper, no. 128, 1979), p. 44.

79. American Insect Control Delegation, *Insect Control,* p. 97.

80. *Guangming Ribao,* February 7, 1980, p. 1.

81. Berner, "Organization and Economy," p. 46; American Insect Control Delegation, *Insect Control,* pp. 101–102.

82. Berner, "Organization and Economy," pp. 49–50.

83. *Renmin Ribao,* September 6, 1980, p. 5; *New York Times,* August 9, 1979, p. 3.

84. *Renmin Ribao,* July 25, 1980, p. 2.

85. Kuo, *Agriculture,* p. 216; American Insect Control Delegation, *Insect Control,* p. 114.

86. American Insect Control Delegation, *Insect Control,* pp. 113–114.

87. Ibid., p. 90; *Renmin Ribao,* July 25, 1980, p. 2.

88. Kuo, *Agriculture*, p. 216.
89. American Insect Control Delegation, *Insect Control*, p. 144.
90. Ibid.
91. *China Reconstructs,* June 1978, p. 35.
92. *China Reconstructs,* October 1978, pp. 36–37; American Insect Control Delegation, *Insect Control*, pp. 133–136.
93. American Insect Control Delegation, *Insect Control*, p. 136.
94. Ibid., p. 151.

Chapter 6

1. For details of the rationing system and the official retail prices of some rationed commodities in Beijing, see Dennis L. Chinn, "Basic Commodity Distribution in the People's Republic of China," *China Quarterly*, no. 84, December 1980, pp. 744–754.
2. Neville Maxwell and Peter Nolan, "The Procurement of Grain," *China Quarterly*, no. 82, June 1980, p. 304.
3. Ibid.
4. *Renmin Ribao*, November 12, 1980, p. 1.
5. *Renmin Ribao*, August 24, 1979, p. 1; February 22, 1980, p. 4; November 23, 1980, p. 2.
6. Nai-Ruenn Chen, "The Theory of Price Formation in Communist China," *China Quarterly*, no. 27, July–September 1966, pp. 36–37; Audrey Donnithorne, *China's Economic System* (New York: Praeger, 1967), pp. 440–441.
7. *Renmin Ribao*, March 19, 1980, p. 2.
8. *Renmin Ribao*, October 13, 1979, p. 2; November 6, 1979, p. 1; June 23, 1980, p. 2; Changsha, Hunan provincial service, October 31, 1979.
9. *Renmin Ribao*, July 24, 1979, p. 3; October 13, 1979, p. 2; March 14, 1980, p. 2. Per capita meat consumption in China was only 11 kilograms in 1979. See *Annual Economic Report of China, 1981* (Beijing: Academy of Social Sciences, 1981), part 6, p. 12.
10. Changsha, Hunan provincial service, November 10, 1979; *Jingji Yanjiu* [Economic research], no. 11, 1980, p. 40.
11. *Renmin Ribao,* January 17, 1982, p. 1.
12. *Guangming Ribao,* November 3, 1979; Guiyang, Guizhou provincial service, November 7, 1979; Guangzhou, Guangdong provincial service, September 12, 1979; *Guangming Ribao*, December 17, 1980, p. 1; January 11, 1981, p. 1.
13. *Renmin Ribao*, September 23, 1979, p. 1; November 6, 1979; October 7, 1979, p. 2; June 23, 1980, p. 2; *Jingji Yanjiu*, no. 11, 1980, p. 38.

14. *Renmin Ribao*, September 23, 1979, p. 1.
15. *Peking Review*, no. 47, November 24, 1978, p. 18.
16. Changsha, Hunan provincial service, November 5, 1979.
17. *Beijing Review*, no. 26, June 29, 1979, p. 11.
18. *Beijing Review*, no. 8, February 23, 1981, p. 8.
19. *China Reconstructs*, January 1981, p. 7.
20. Donnithorne, *Economic System,* pp. 337–346; Ronald Berger, "Financial Aspect of Chinese Planning," *Bulletin of Concerned Asian Scholars,* vol. 6 (1974), p. 16.
21. *Renmin Ribao*, September 6, 1980, p. 4.
22. *Renmin Ribao*, February 10, 1979, p. 1.
23. *Renmin Ribao*, September 25, 1979, p. 2.
24. *Jingji Yanjiu*, no. 2, 1980, p. 12.
25. Ibid., p. 13.
26. Ibid.
27. Ibid., p. 15.
28. Note, however, that besides agricultural production, the state farms also have other functions. Because they are generally located in the border areas, their development has the defense significance of populating and developing the border areas.
29. *Jingji Yanjiu*, no. 2, 1980, p. 14.
30. *Renmin Ribao*, September 1, 1980, p. 1.
31. *Beijing Review*, no. 49, December 7, 1979; *Renmin Ribao*, September 2, 1980, p. 1; November 27, 1980, p. 2.
32. *Peking Review*, no. 50, December 15, 1972, p. 17.
33. Credit cooperatives were established in the early 1950s as independent rural cooperatives to serve members' needs. They were absorbed into the production brigades in 1959. In 1962 they were placed under the leadership of the banks. See *Renmin Ribao*, December 4, 1980, p. 5.
34. *Renmin Ribao*, May 11, 1979, p. 2.
35. *Renmin Ribao*, June 30, 1980, p. 3.
36. *Renmin Ribao*, March 9, 1979, p. 3.
37. Ibid.
38. Ibid.
39. *Renmin Ribao*, October 29, 1979, p. 2.
40. *Renmin Ribao*, September 8, 1979, p. 3.
41. Public accumulation funds also include a small welfare fund for cooperative medical services, subsidies to poor households, and other welfare expenses. However, the welfare fund constitutes a very small percentage of public accumulation funds.
42. See Azizur Rahman Khan and Ng Gek-boo, "Achievements and Incentives in Communal Agriculture: The Case of China," in Dharam

Ghai, Azizur Rahman Khan, Eddy Lee, Samir Radwan (eds.), *Agrarian Systems and Rural Development* (New York: Holmes & Meier Publishers, 1979), pp. 260-261.

43. Ragnar Nurkse, *Problems of Capital Formation in Underdeveloped Countries* (Oxford, England: Oxford University Press, 1953), pp. 36-47.

44. Guangzhou, Guangdong provincial service, August 23, 1979.

45. Thomas G. Rawski, "Choice of Technology and Technological Innovation in China's Economic Development," in Robert F. Dernberger (ed.), *China's Development Experience in Comparative Perspective* (Cambridge, Mass.: Harvard University Press, 1980), p. 210. See also American Rural Industry Delegation, *Rural Small-Scale Industry in the People's Republic of China* (Berkeley: University of California Press, 1977), pp. 63-110 for technical details.

46. New China News Agency, July 10, 1979.

47. *Renmin Ribao*, September 10, 1979, p. 1.

48. *Renmin Ribao*, October 19, 1979, p. 2.

49. *Renmin Ribao*, September 10, 1979, p. 1.

50. Shenyang, Liaoning provincial service, September 6, 1979.

51. *Renmin Ribao*, August 13, 1975, p. 3; April 12, 1979, p. 2.

52. Ibid.

53. *Renmin Ribao*, August 18, 1979, p. 2.

54. *Renmin Ribao*, April 12, 1979, p. 2.

55. *Renmin Ribao*, January 18, 1979, p. 2.

56. *Renmin Ribao*, June 25, 1980, p. 2; December 4, 1980, p. 2.

57. For the text of the new regulations, see *Renmin Ribao*, May 16, 1981, p. 4.

Chapter 7

1. The World Bank, *World Development Report, 1980* (New York: Oxford University Press, 1980), p. 143; John S. Aird, "Population Growth in the People's Republic of China," in Joint Economic Committee, *Chinese Economy Post-Mao* (Washington, D.C.: U.S. Government Printing Office, 1978), p. 467.

2. Calculated from Table 3.1.

3. Kang Chao, *Agricultural Production in Communist China, 1949-1965* (Madison: University of Wisconsin Press, 1970), pp. 130-135.

4. *Renmin Ribao,* November 30, 1979, p. 3. *Guangming Ribao,* December 19, 1980, p. 4, reports that, in Sichuan, rainfall in 46 percent of the counties declined by 15-20 percent in the 1970s because of deforestation.

5. *Renmin Ribao*, August 19, 1981, p. 1; September 3, 1981, p. 2; October 31, 1981, p. 2.

6. *Beijing Review*, no. 45, November 9, 1979.
7. *Beijing Review*, no. 48, November 30, 1981, pp. 16–17.
8. *Renmin Ribao*, September 18, 1980, p. 2.
9. Ibid.
10. *Renmin Ribao*, February 2, 1979, p. 2.
11. *Beijing Review*, No. 45, November 9, 1979, p. 6.
12. *Renmin Ribao*, October 8, 1981, p. 2.
13. This is the major reason why China is now permitting urban youth to establish their own collective or private business.

Index